The Political Theories of Risk Analysis

The International Library of Environmental, Agricultural and Food Ethics

VOLUME 4

THE POLITICAL THEORIES OF RISK ANALYSIS

by

Nicholas P. Guehlstorf

Southern Illinois University,
Edwardsville, IL, U.S.A.

Springer

A C.I.P. Catalogue record for this book is available from the Library of Congress.

ISBN 1-4020-2881-4 (HB)
ISBN 1-4020-2882-2 (e-book)

Published by Springer,
P.O. Box 17, 3300 AA Dordrecht, The Netherlands.

Sold and distributed in North, Central and South America
by Springer,
101 Philip Drive, Norwell, MA 02061, U.S.A.

In all other countries, sold and distributed
by Springer,
P.O. Box 322, 3300 AH Dordrecht, The Netherlands.

Printed on acid-free paper

springeronline.com

Printed in the Netherlands.

CONTENTS

ACKNOWLEDGEMENTS

As recognition of the pleasurable, necessary, and beneficial gifts for which my scholarship in this book is indebted, I am beholden to many professors and family members. To Dr. Penny Weiss, your total and unremitting faith in me has allowed and forced me to complete a project that is respectable political theory and good social science. To Dr. Robert Bartlett, I appreciate all of the time you have spent on my work and will always be indebted to you for supporting my ambitious scholarly attempt to wed American environmental policy with liberal political theory. And finally, to Professor Paul Thompson who helped me formulate my immature ideas about environmental risk into a matrix of ethics and policy, which I have since attempted to present in a coherent book. I am grateful that you have never judged the success or failure of every decision I have made in developing my modest thoughts because my scholarship involves much of your own publications, personal contacts, and governmental papers. Although I am appreciative of you for presenting me with an endless list of scholars who can speak to my work, I am more thankful to you for providing me the freedom necessary to acquire my own voice. A variety of people at and reviewers for Kluwer Academic Publishers have helped develop the content and tone of this book and I am forever grateful for the feedback and opportunity provided to me through the Humanities division and the series International Library of Environmental, Agricultural, and Food Ethics. Additionally, I would like to recognize Dr. Owen C. Phelps who edited my original manuscript. Thank you for always treating my work and me with more respect than either deserved or earned. My most profound gratitude is extended to my best friend, only partner, and first proof-reader, Kristin. Finding you has helped me to realize all of my dreams.

CHAPTER ONE

RISK, RATIONALITY AND LIBERALISM

1.INTRODUCTION

Most scientific researchers, American governmental regulators, and principal business leaders maintain that there is very little risk and a myriad of benefits in genetically engineering food. They argue that biotechnological methods of splicing and transferring genes for the production of food could result in enhanced farming profits, less environmental contamination, and new agricultural products that are valued by consumers. According to the United States Department of Agriculture (USDA), 68 percent of the soybeans, 69 percent of the cotton, and 26 percent of the corn harvested in 2001 were genetically altered (PEW, 2003). However, notwithstanding its existence in the American food supply, convincing arguments by producers, and data offered by governmental agencies, agricultural biotechnology has its critics. While there is no scientific evidence that genetically engineered food is unsafe to eat, some people worry about the potential for unknown allergens, the increase in natural toxic substances, the decrease in nutritional value, and unintended environmental consequences (Burros, 1999b). Environmental concerns in particular have led to the illegal destruction of genetically engineered crops by protestors, the voluntary rejection of gene-altered ingredients by certain companies, and the discussion of reforming the regulatory approval process by governmental agencies (Barboza, 1999a).

According to governmental and industry experts, such actions are seemingly irrational because all genetically modified food sold commercially has passed scientific risk analyses for human and environmental health (P. Thompson, 1998b). The fundamental problem for concerned citizens, however, is that they see political decisions about the risks associated with food biotechnology conducted without consideration of their values, consent, or deliberation. In order to offer some thoughtful insight into current public policy debates about modern agricultural methods, this book explains the administrative practices of environmental risk analysis in terms of their foundational political theories of societal risk. Such a philosophical examination will offer some justification for the dominant paradigm of rationalism used by public administrations and might also give some validation to the structure of liberalism used by environmentally aware citizens. As such, this work will categorize all pertinent risk studies as using either a rational framework or a structural design. Within that conversation, I hope to define risk, rationality, and liberalism as clearly as possible, as they are the three most important concepts of this book. A categorization of applicable research follows, after which the reader will find a concise academic discussion of American environmental policymaking, a brief analysis of the agricultural debate regarding genetically modified organisms, and the introduction of the case study of the biotechnology used in milk production. While the objective of this book is to theoretically describe the correlation of environmental risk practices and their current underpinning political philosophies, some prescriptive insights about the practical relationship of environmental risk theory and practice might be drawn for use with issues that arise in the future.

2.STATEMENT OF THE RESEARCH PROBLEM

Risk analysis is the process of making societal choices about ecological dangers and human health hazards by determining the probability of harm using scientific techniques of evaluation. The analysis of environmental risk—the rationally determined likelihood and magnitude of an adverse event—is a complicated procedure that quantitatively discerns acceptable levels of hazard while working with incomplete environmental information, limited scientific resources, and multiple external influences (Armour, 1997). According to the technical risk analysts, however, this is not problematic because the procedures they use to scientifically estimate risk are kept separate from considerations of the political acceptability of risk and the public understanding of risk.

Risk analysis consists of: (1) the risk assessment that includes the identification of hazards, exposures and characterization of risk; (2) risk management that classifies, selects, and mitigates measures used to control or reduce risk; and (3) risk communication or consensus building of interests among risk experts, policymakers, and affected segments of the public (Boroush, 1998). Despite the fact that risk analysis aimed at achieving objective, apolitical, and rational decisions has become a specialty in some bureaucratic settings, many scholars in risk studies have concluded that risk statements are, in fact, subjective performances of differing judgments (Douglas & Wildavsky, 1982; Mayo, 1997; Whipple, 1992). That is, the benefit-risk analysis—that may entail the assessment, management, and communication of a new technology—only appears to be value free with its technical language that discusses the potential of uncertain hazards by defining risk as the "average annual probability of failure". In fact, not only do such quantitative discussions of risk usually favour production, development, and commerce, they also often attempt to exclude qualitative factors like equity, welfare, and democracy.

The problem of interpreting risk analysis as a technical-rational project, and not a social-political process, is further complicated by the elite power of bureaucratic decisionmaking in a liberal political society (Stone, 1997). Discerning the proper amount of democratic control, however, becomes troublesome because the authoritative mechanism by which risk is defined and the technological areas in which it is involved are not well understood by the public (Gillroy, 1993). For instance, although food biotechnology—the direct genetic modifications of microbe, plant, and animal deoxyribonucleic acid (DNA) in order to increase profit, improve breeding, or intensify traits desired by consumers—affects everyone in the United States, few Americans have ever evaluated the costs and benefits of modern agriculture, individually consented to eating biotech foods, or collectively discussed the societal risks of their use (Westra, 1998). Instead, most Americans either fail to know about genetically altered food or innocently trust that there are no questions about bioengineered products that have passed all governmental safety studies and risk analyses. Governmental agencies and other scientific regulators of biotechnology communicate using the elite linguistics of science, law, and economics instead of the public languages of equity, power, and trust.

As a consequence of the divorce of bureaucratic practice from social theory, environmental risk is traditionally understood as a technical judgment that seems to dismiss political justification, not need social permission, and eliminate public discourse (Beck, 1995). It is a process that puts democracy at risk because standard risk evaluations that emphasize quantitative techniques fail to address the qualitative causes of public concern about environmental problems (Westra, 1998). This is important because lay assessments of risk often reveal what dangers society regards as acceptable and what hazards spur fear, apprehension, and protest (K. S. Shrader-Frechette, 1991). Furthermore, it has been empirically proven that the public perceptions of risk not only transcend technical analysis, but also in some ways are more sophisticated because the public has a greater consideration of the values and concerns ignored in the statistical conclusions of technical elites (Slovic, 1992). Because expert judgments have a subjective bias toward having the highest utility of outcomes for production, industry, and commerce, experts might not be making decisions that consider: general public perceptions, opinions of the minority, outcomes that have a low probability but extremely high consequences, and possible negative long-term effects of certain short-term benefits.

The two positions in the philosophical debate on risk—that is, those held by specific experts and the general public—have been characterized as being positivist/relativist, scientific/constructivist, or probabilist/contextualist by different risk scholars (Horning, 1993; K. S. Shrader-Frechette, 1991; P. B. Thompson & Dean, 1996). While not all identical, each dichotomy offers a similar schema for mapping out alternative conceptions of risk. Furthermore, while there are viewpoints in the literature on risk that fall along the spectrum, most views are weak versions of the extreme poles (P. B. Thompson & Dean, 1996). At one extreme is the rational-scientific conception that maintains that risk is best defined by the objective probability of a hazardous event. This positivistic, scientific, and probabilistic position is held by governmental and industry technicians and is best explained by Chauncey Starr and Chris Whipple in their classic essay "Risks of Risk Decisions". The polar opposite is the liberal-social understanding best explained by Mary Douglas and Aaron Wildavsky in *Risk and Culture* and is represented by environmentalists who oppose many modern American agricultural processes. Advocates of this position have been called relativists, constructivists, and contextualists because they maintain that risks are best defined by individuals whom are at risk and therefore reflect subjective notions like beliefs, emotions, and culture.

The leading approaches in risk studies are either to persuade the general public to be more rational in its understanding of risk, or convince the technical elite to become more democratic in their evaluation of risk. The problem with both of these trends in the social studies of risk is that they neglect the idea that although technical elites and the general public are arguing about risk in radically dissimilar political languages, they are both operating from similar political philosophies. That is, most scholarly attempts to connect the current divide of risk erroneously assume that one position can only be validated with the negation of the other. Traditional policy analysis incorrectly finds that technical and social communities have more disagreements than agreements. Therefore, this book focuses on identifying the rational practices inherent in the expert knowledge of risk, explaining the social theories that undergird the risk perceptions of the public, and discussing the limitations of both the practice of science and the social theories of risk. The purpose of this three-fold process is to allow each seemingly opposed political faction to understand and consider the other faction's theoretical or normative position.

Any thorough review of conceptions of risk would discern that very few policy scholars or planners maintain either an intensely rational or extremely social understanding of risk, arguing that either the technical definition or the public perception of risk is universally applicable. Without unqualified confidence and certainty, however, planners and scholars operate within the dichotomous paradigms of risk because, despite their diffidence, they need to make a political or professional decision that attempts to avoid harm according to some respected standards.[1] These seemingly opposing paradigms are not diametrically opposed, inasmuch as both are attempting to achieve the same objective of avoiding harm. I propose that they are operating from the same political foundation of liberalism—a philosophy based on the belief in progress, the essential goodness of humanity, the autonomy of the individual, and the social organization of civil liberties and securities. Going beyond the current technical and public dichotomy of risk analysis therefore requires an understanding of the core liberal foundation of their political separation. Quite simply, in order to understand their differences, I will examine their sameness. The purpose of identifying the central political values of risk analysis is not to idealistically combine the opposing interests of scientific experts and lay citizens or even persuade one group to accept the other; rather it is to create a political space where each political faction can inclusively understand the opposing group's issues and values in order to communicate.

It is not overly important here to differentiate the theoretical nuance of modern liberalism from that of classical liberalism because both the rational-scientific practice and the liberal-social theory of risk are generally concerned with modern liberalism. Classical liberalism, beginning in the seventeenth century, is based upon a tradition of natural law that formulates for people a private sphere within which they can partially determine their own destiny while being shielded from social control (Wiser, 1983). Classical liberalism has a strict regime of private property rights that defends the market economy, claiming that any redistributive policies of the state violate the fundamental individual rights that are the basis for liberalism. Modern liberalism, however, supports a considerable design of civil liberties and economic securities such as welfare, education, environmental protection, and health care (Wiser, 1983). Modern liberalism—which does not impose a universal approach to life, but leaves its citizens as free as possible to choose their own values while respecting a similar liberty for others—replaces the moral basis of natural rights with either a utilitarian or a neutral framework but still accepts classical liberals because it does not have a bias for a preferred way of life.[2]

[1] For example, even the most liberal scientists and technicians who acknowledge that their assumptions on how to collect data and interpret results are not value-free, at the end of the day, find themselves in the scientific-rational camp because they privilege peer reviewed conclusions over the tragically biased and subjective ideas of the public. Equally, Kristin Shrader-Frechette's philosophical middle position of "scientific rationalism" despite its attempt of respecting rational communities finds, its locus in the social-liberal camp because in the end she wants to mediate ethical conflicts between experts and laypeople by democratizing rational risk decisions

[2] Sandel (1984) argues that the utilitarian view as explained in John Stuart Mill's *On Liberty* states that, "The only freedom which deserves the name [liberalism]. .does not depend on any notion of abstract right, only on the principle of the greatest good for the greatest number" (2). According to Sandel, most contemporary political philosophers consider utilitarianism an inadequate foundation for liberalism because the utilitarian majority fails to secure the liberal promise of not imposing values upon individuals.

Notwithstanding public opinion, the core of most debates about contested agriculture and risk policy in modern farming is not between classical liberalism and modern liberalism; it is about modern liberalism finding its ecological rationalism or realizing its sustainable limits (Ophuls & Boyan, 1992). More importantly, some of the environmental problems associated with modern liberalism can be discerned by investigating the political theory that provides a foundation for risk analysis. This is true because from the view of modern liberalism, there is no scientific certainty or normative truth supporting an absolute definition of risk. Instead, the interpretation of risk is contingent upon both its social and scientific context. The hazardous reality of a proposed action is defined for and by a community of interested parties and relevant practitioners (P. B. Thompson & Dean, 1996).

3.ENVIRONMENTAL POLICY AND ANALYTICAL FRAMEWORK

Many political scientists who study environmental policy have argued that environmental decisionmaking consists of scientific knowledge, specific technology, complicated economics, and political ethics (Dryzek, 1997; Kraft, 1996; Paehlke, 1990). Given the specific knowledge of environmental policy today, however, most ethical concerns about the political values of democracy are overshadowed by technology, science, and economics. Risk analysis is therefore undemocratic because bureaucratic, scientific, and industry experts control the public policy process of American environmentalism. That is, the American administrative state has created a rationalized bureaucratic society that does not include the complete performance of the following political processes of risk analysis: (1) a liberal or utilitarian justification; (2) a liberal consent or social contract; and (3) a liberal debate or critical social theory of public discourse. Although utilitarianism, social contract, and critical social theory are found in the foundational premises of risk analysis, none of these theories are executed to their full capability.

Furthermore, as massive environmental risks continue to threaten Americans, policy debates—although scientifically, legally, and technically complex—need to be perceived as not being politically ambiguous or beyond public understanding, consent, and deliberation. Rational practices of risk analysis are susceptible to explanation by liberal political theories because participants in the political processes of risk analysis reasonably

---

For instance, Immanuel Kant in *Groundwork of the Metaphysics of Morals* argues that all empirical justifications, like utility, are unfit to serve as a moral law in a political philosophy. More importantly, writes Kant, the utilitarian principle obliterates human dignity by treating people as means to the happiness of others and not as ends in themselves, worthy of respect. The neutral framework of modern liberalism, first formulated by Immanuel Kant, is a justification based on the distinction between rights and liberties, and the autonomous choices of the good that people may choose to pursue within the framework.

negotiate, refine, and bargain with the policy stakeholders interested in the assessment, management, and communication of societal risk. This book, therefore, investigates risk analysis by qualitatively analysing the policy debates about food biotechnology between the rationalism of public administrators and the liberalism of the general public.

Policy studies have traditionally been oriented towards a rational design, with economics and many of its applied methodologies serving as the dominant approaches (McCool, 1995). The three frameworks of risk practice are: (1) rational choice theory in the assessment of risk; (2) administrative decisionmaking in risk management; and (3) institutional participation in the communication of risk. All three of these rational approaches are, however, inadequate for the liberal consideration of all risk positions in society and therefore too limited to prescribe alone complete policy remedies that inclusively understand environmental values. Instead of approaching risk policy as rational "experiments," this analysis considers them as political "arguments" about liberalism because methodologically they are social interactions among democratic policymakers, scientists, and citizens (Fisher & Forester, 1993; Majone, 1989). That is, I consider risk analysis not as an internal discussion of scientific inquiry aimed at objective truths but an external conversation of political debate aimed at subjective realities. For instance, the acceleration toward fewer and larger farms with much specialization is best understood as a utilitarian political decision based on the economic magnitude of potential benefits rather than the scientific effect of biotechnology on contemporary agriculture. Whereas the former is a controllable decision among a liberal collective, the latter is a physically determined technical consequence.

As a consequence of considering risk analysis as a political exercise, this book analyses risk practices about contested agricultural technologies as arguments based in liberal political theories. The premises of my research, illustrated in Table 1, are that: (1) risk assessment is the practice of rational choice which is an argument for utilitarianism; (2) the theory of social contract is the argument practiced in the administrative decisionmaking of risk management; and (3) the practice of institutional participation in risk communication is an argument for critical social theory. The analytical framework of my work is as follows:

Table 1. Risk Analysis

	Risk Assessment	*Risk Management*	*Risk Communication*
Rational Practice	Rational Choice	Administrative Decisionmaking	Institutional Participation
Liberal Theory	Utilitarianism	Social Contract	Critical Theory

Risk analysis is the rational process of making political choices about ecological dangers and human health hazards by determining harm without reasonably evaluating the utility, contract, or critical discourse of a liberal society. Therefore, the current rational practice assumes that the majority of citizens in the United States will accept the utility of risk evaluations, agree with the power in risk management decisions, and trust governmental and industry scientists when they communicate that they are guarding society against environmental and human health risks. As mentioned, it appears from recent political events and scholarship that many Americans have reservations about these assumptions.

4.CASE STUDY DESIGN

The specific question investigated in this book is: "Can a liberal democracy, in theory and practice, adequately respond to the specific ecological challenges of risk associated with new, emerging, and contested agricultural technologies in modern food production?" As a consequence of the volume of environmental policies that utilize risk analysis in agricultural biotechnology, I focus my attention on the case of the bovine growth hormone used by the American dairy industry. This research will explore the political theories that ought to inform the political practices of risk by qualitatively examining the language used in primary industry and governmental documents and the secondary research regarding risk assessment, management, and communication. American agriculture is the most serviceable area of study for my project because it has a history of environmental criticism, it is regulated by multiple governmental agencies, it employs some of the newest and most contested technologies, and it currently involves a great deal of political discourse among policy planners, industry technicians, and the public (P. Thompson, 1995). Finally, as a consequence of the number of people affected by and concerned about the human health and environmental dangers of food, my conclusions can be generalized to many contemporary practices of risk analysis (Marshall & Rossman, 1989).

In order to offer some clarity about how rational decisionmaking functions in liberal risk analysis, the scientific, economic, and social dynamics that contributed to the acceptance of recombinant Bovine Somatotropin (rBST) in American agricultural policy will be investigated. Analysing the most popular, researched, and controversial case of food biotechnology—the synthetic growth hormone injected into dairy cattle to stimulate milk production—may illuminate some of the ethical and political problems of recent biotechnological developments in the United States (Gendel, Kline, Warren, & Yates, 1990; Lehman, 1995; Powell & Leiss, 1997; Ratzan, 1998; P. Thompson, 1998a). That is, the development of a comprehensive theory of risk analysis for the political issues associated with rBST could be practiced for most other food biotechnologies (Mephan, Tucker, & Wiseman, 1995). It should be noted from the beginning that rBST is chosen as the central case study of this book because it illustrates well in three ways the politics of risk regarding agriculture biotechnology.

First, it established the current American regulatory framework for all food biotechnology. Mainly that the potential risks of genetically engineered plants and animals are examined by industry and approved products are not governmentally labelled. Second, it created the groundwork for the scientific and legal arguments for the factional debates between industry and environmental groups for all future products and processes of biotechnology. Third, it was the "stand-in" or "symbolic" case study of industrial agriculture debated in the United States. Regardless of the fact few scholars saw any inherent environmental impacts resulting from rBST use, it became the locus for talking about all the issues of ecological harm and food safety that may result from modern farming. Although there did not appear to be any serious safety problems with rBST, it was the first publicly available genetic technology and the regulatory dispute taught opponents of biotech that environmental risk is about human danger and it gave advocates of biotechnology the lesson that technological, scientific, and economic benefits must have some redeeming social value (Charles, 2001). That is, although rBST was unnecessary because dairy farmers were already producing too much milk and unpopular because it made one of Nature's most natural substances synthetic it was governmentally accepted because the genetically engineered milk was not dangerous, harmful, or risky—"the probability of future loss was low" (Byrd & Cothern, 2000, 2).

Researching the underpinning liberal political theories of the rational practices associated with the risk analysis for rBST therefore has significant implications for the contentious scholarly and public debate regarding most agricultural risk issues. More importantly, this analysis will explain main contexts of agreement and disagreement between the rationalistic problem-solving model of practice used in environmental policymaking and the liberal politicised-determinant structure of theory (Fisher, 1990; Fisher & Forester, 1993). Understanding the link and the disconnect between the rationalism of the regulatory state and the liberalism of risk society are the main political and cultural concerns of advanced modernity (Hood, Rothstein, & Baldwin, 2001, 4).

Bovine Somatotropin (BST) is a naturally occurring protein produced by the pituitary gland in cattle that promotes growth in calves and milk production in lactating cows. Administering pituitary extracts from slaughtered cows to lactating animals in order to increase milk production is an American farming practice that dates back to the 1920s (Monsanto, 2001a). The veterinary drug recombinant bovine somatotropin (rBST) also referred to as recombinant bovine growth hormone (rBGH or BGH) is produced through biotechnology and when administered to cows can increase milk production and length of lactation cycle by ten to fifteen percent (FDA, 1999a). In 1979 the American based company Monsanto obtained, from the biotech firm Genentech, the rights to fully develop and commercialise the recombinant BST process, which is a molecular biology procedure of genetic splicing, first done to produce human insulin. To make rBST (trade name Posilac), Monsanto takes the gene that produces BST in the pituitary gland of a cow and splices the DNA gene sequence that carries the instructions for making BST into an *E. coli* bacterium. The genetically engineered organism is then grown under classical fermentation conditions, the *E. coli* is killed, all unnecessary proteins and membranes are separated from the BST, and it is finally incorporated into a pharmaceutical drug to be inserted into lactating cows (Monsanto, 2001a).

It has been scientifically determined that rBST is virtually indistinguishable from BST, that both BST and rBST are largely destroyed by pasteurisation, that humans are not responsive to growth hormones obtained from cows, and that treating cows with rBST does not affect the nutritional qualities of milk (HCPD, 1998). Monsanto has been evaluating Posilac since 1981, and the American government started reviewing the human health effects, food safety, product efficacy, animal safety, manufacturing quality, and environmental impact in 1985 (Monsanto, 2001a). Almost all accredited members of the medical, veterinary, and dietetic communities assert that there are no significant human health risks or food safety problems connected with any products associated with the use of the biotechnology drug rBST in the dairy industry. Using rBST to treat cows in the United States was approved by the Food and Drug Administration (FDA) in February 1994, and since then the practice has been extensive. Similar regulatory conclusions with respect to food safety have been reached in 34 other countries (IFST, 1999).

The aforesaid data notwithstanding, many countries with well-developed regulatory agencies, including Canada and several in Europe, have not approved rBST for dairy production. Expressed concerns have focused on animal health, labelling, and socio-economic consequences. Similarly, during the nine year approval process in the United States, lay opponents of rBST claimed that human health questions about the safety of milk produced by rBST cows were not resolved, that risks associated with the well-being of the cows treated with unfamiliar biotechnology were not understood, and that social and economic consequences of the new technology on the culture of the American family farm were not predictable. Finally, critics of rBST contend that some public health risk assessments—both characterization and exposure—have indirectly pointed to an association between immune growth factor one (IGF-I) levels, the hormone used to stimulate a cows' gland, and an increased relative risk of certain human cancers.

Organizations and individuals opposed to modern biotechnology and agriculture have targeted rBST because of the alleged trade-motivated regulatory rules of the American government, as portrayed in some media stories. Additionally, there was a lot of criticism of the drug, agriculture, and biotechnology company, Monsanto, that has the sole governmental approval and patent to market rBST in the United States. Aside from general criticisms that it is a multinational company that exercises hegemonic influences on health, farming, and environmental policy, it is condemned for implementing the current regulatory structure of food biotechnology. That is, most current public policy debate is focused on reforming the general regulatory structure for genetically modified organisms (GMO) used in modern agriculture and food production. Specifically, critics contend that the FDA should establish criteria for a "GM Free" or "non-GM ingredient" food label and that internal governmental testing of all GM food products should replace the existing "industry examination with FDA consultation-safety review" (FMI, 2001). It is my argument that rBST opposition in the U.S ten years ago, like opposition to genetically modified food at the global level today, is less about the human risks or ecological safety and more about how the technology is used, sold, marketed, and communicated (Pistorius & Wifk, 1999).

5.IMPORTANCE OF THE RESEARCH

No study of environmental risk policy includes any sustained examination of how political theory applies or ought to apply to the process. This omission is noteworthy because political theory can be a topic for understanding the evaluations made in natural resource policymaking as well as a methodology for discerning the decisionmaking processes in an environmental political administration. As a consequence of modern philosophy being constructed from both empirical and sociological traditions, it can autonomously describe the past and sketch the future technological society (Keulartz, Korthals, Schrmer, & Swierstra, 2002). This normative omission of risk also is unexpected given the quantity and scope of political philosophy that has influenced or could impact environmental problems, politics, and policy. In contrast, this study addresses the defects in the current rational-scientific reasoning for risk analysis and, as an alternative, proposes a normative-liberal understanding of risk analysis. The rudimentary foundation of political liberalism is the idea that citizens are separate individuals, each with their own aims, interests and conceptions of the good who, therefore, seek a framework of civil rights that will enable them to realize their capacity as free agents, consistent with a similar liberty for all others in the polity (Sandel, 1984). Considering American public policy as an application of this liberal political philosophy is not a radical concept (Ingram & Rathgeb, 1995). Although there are a few articles in the risk literature that address some of the relevant philosophical work in American environmental public policy, there is no comprehensive study of modern political theory and the contemporary practice of risk analysis.

In fact, many political scholars maintain that United States public policy could be the ideal academic intersection between liberal political theory and practice (W. Dunn, 1983). Therefore, liberal political philosophy can be seen as a series of policy alternatives trying to establish the best social structure of liberalism. It is my hypothesis that if governmental practices of risk clarify the liberal goals of its bureaucratic agencies, American public policy might be able to interpret correctly and react appropriately to the risks it actually encounters with respect to the environment. Furthermore, defining the liberal theoretical objectives of risk policy will allow bureaucracies to come to terms with the real limits of their administrative practices. The merit of this case study of agricultural risk lies in the clarity of my descriptions of the complex association of political theory, public policy, and public administration. This complex association helps illustrate problems inherent in the dominant perspective of risk analysis that have incorrectly dismissed significant questions of liberalism in environmental policy.

Political philosophy is interested in how to construct the "good" or "just" society and problems regarding how public life ought to be organized (Strauss, 1959). One of the purposes of political theory research within the discipline of political science is to uncover the view of the human condition implicit in each government action. Governmental processes, policies, and programs cannot be sufficiently understood without an awareness of their vital political assumptions. Studying the relationship between political philosophy and governmental action serves to heighten this awareness. A theoretical investigation of the practice of regulatory risk regimes can critically articulate the theoretical problems of current risk analysis (Hood et al., 2001). Although there are numerous political, social, and economic critiques of risk analysis, there is no comprehensive political theory scholarship of risk. In examining the discourse of policy planners both with themselves and with the public, this book will expand the theoretical understanding of administrative risk deliberations and help to explain some of the difficulties democracy encounters with American environmentalism.

6.CHAPTER OUTLINE

In the next chapter, "The Technical Values in the Expert Judgment of Risk," the FDA's empirical efforts to define the probability of societal harms related to the dairy industry's use of the genetically engineered bovine growth hormone, rBST will be presented and analysed. This section of the book will carefully consider the value judgments made in defining the issues as well as who benefits and how much is at risk in modern agrarianism. The focus of this chapter will be on the arguments made by the proponents of new biotechnological methods in agriculture. The economic effect of biotechnology on agriculture and the acceleration toward fewer and larger farms with much specialization will be discussed. The political impact the biotechnology industry has had in governmental lobbying in the United States over the past two decades will also be addressed. The argument that rational cost-risk evaluation is objective science will be critiqued, particularly with regard to the value-laden perspectives found in its technical rules, strategies, and methodologies. Rather than being a value-free, technical, and rational assessment, cost-risk evaluation will be exposed as a utilitarian political decision rule.

Chapter Three, "The Utilitarian Assessment: the Politics of Living with the Smallest Risk for the Greatest Number" examines the gap between the modern theory of utilitarianism and the contemporary practice of rational choice in risk assessment. Jeremy Bentham's *The Principles of Morals and Legislation* is used as an outline for a comprehensive theory of public policy. The utilitarian procedure for making law and policy—to project the consequences caused by the proposal and its alternatives—is easily compared to the rational choice methodology of identifying the optimal choice. Initially, I question whether utilitarianism offers a comprehensive structure for liberal principles, including respect for rights of the human condition and environmental integrity. Secondly, I focus special attention on the challenges posed by using only rational choice logic in democratic decisionmaking.

Next, I reason that environmental risk assessments based primarily on Bentham's utilitarianism are a process of subjective rationality that politically distorts the liberalism desired in John Stuart Mill's *On Liberty*. Finally, using feminist standpoint epistemology, I critique Mill's failed attempt at objectivity and argue that risk issues associated with agricultural biotechnology are a prominent demonstration of the defects of using only a rational choice methodology for defining ecological harm. That is, while rational choice provides a "probabilistic assessment of risk" in food biotechnology that is helpful in determining a decision rule, that must not be the only criteria used in accepting or rejecting a risk in American environmental public policy.

The organizational function of risk in the history of modern political philosophy is explored in Chapter Four, "The Practice of Risk Management and the Theory of Social Contract." Within this investigation, I assert that risk analysis in American environmental policymaking is a theoretical construction of social discourse that is practiced by the estimation of the scientific method. Therefore, the general purpose of this chapter is to question if risk management presupposes any particular political consent, or functions by disregarding liberal values. Specifically, recent risk evaluations in food biotechnology are examined from the perspectives of social contract theory and the practices of administrative rational decisionmaking. I argue that although risk management trades on John Rawls's social contract philosophy to justify political authority, it fails to incorporate the notion of individual consent from John Locke and the idea of public sanction from Thomas Hobbes. Finally, the feminist critique of social contract is suggested because it allows for a more inclusive approach to consent-based decisionmaking.

Chapter Five, "A Theoretical Case for Reasonableness" explains environmental activists' concerns with the marketing, inspection, and regulation of recombinant DNA gene transfer techniques in milk production. This case study provides a perfect example of the ethical concerns the public has with rational choice calculations of personal harm and societal risk. This chapter explains why some citizens fail to believe that genetically modified organisms will positively revolutionize farming, as risk managers and agricultural experts contend. It also explores why these transgenic protestors argue that new food biotechnologies could destroy the ecosystem and, although there might be benefits for agribusinesses, the risks for others and the environment are too great to justify their use.

The philosophical foundations of populist reform and environmental civil disobedience will be analysed. Although the position of transgenic protestors is as one dimensional as that of industry spokespersons, we will see that their politics of disapproval are the by-product of undemocratic risk management. Thus, we will analyse how the same scientific, economic, and social dynamics that contributed to the legal acceptance of rBST fanned the flames of suspicion and resistance to new food biotechnologies. Investigating the political opposition to rBST will clarify some of the ethical problems of other biotechnological developments in the United States. More importantly, I critique administrative decisionmaking in risk management for not attempting to be democratic or comprehensive.

In Chapter Six, "The Risks of Environmental Democracy," I argue that because risk management does not reflect a complete social contract theory and is therefore unable to generate an inclusive conception of risk, risk communication currently fails to adequately explain contemporary debates over technology. This chapter seeks to discern whether the political philosophies found in critical social theory can inform the way policymakers and public administrators communicate the risks involved in food biotechnology. Environmental democracy has been theoretically defined as governance that meaningfully involves the public in environmental policymaking. Risk communication—the consensus building of interests among risk experts, policymakers and affected segments of the public—could be the ideal practice of environmental democracy. Because environmental risk has been traditionally understood as a technical judgment regarding the statistical probability of a hazardous event and its perilous consequences occurring, it often eliminates public discourse.

Interpreting risk analysis as a political process rather than just a technical project can alter the dominant understanding of environmental risk. This chapter suggests that environmental public policy would be well served by drawing upon the critical social theories of democracy, both deliberative and communicative, in order to publicly debate accepted levels of risk in society. The performance of a political community that discursively learns from expert interpretations of harm and public perceptions of risks originates in the early work on language by Jurgen Habermas and is further expanded by the practical importance of the environment exposed by Herbert Marcuse and Ulrich Beck.

In Chapter Seven, "Liberal Communication Needed for Rational Practices," the issue examined is the use of institutional discourse among industry technicians and administrators in the Environmental Protection Agency (EPA), USDA, and FDA. At stake in this poorly communicated risk analysis are billions of dollars in investments by the biotechnology industry and American farmers who have rapidly adopted GMO versions of milk, corn, soybeans and cotton to produce higher yields or resist pests. Because the American bureaucratic culture did not learn any lessons from the dairy debate, the divorce of rational risk practice from liberal risk theory in developing public policy continues today. The lack of public discourse about the role of transgenic foods in the future of agriculture is a vivid example of the need for a broader debate of risk—one that inspires citizen trust and reflects democratic values. Because risk is laden with ethical implications for the entire society, its political analysis must become part of democratic life in America.

This chapter and the book conclude by suggesting that risk analysis should be restructured as a reasonable political rationalization of both empirical science and normative theory. In the end, decisionmaking on environmental issues should occur only after societal risks have been understood, evaluated, and prioritised by a democratic community. It is my suggestion that this can only occur if both the technical experts and the concerned public realize and appreciate the dissimilar political values they have with regard to risk. This value matrix, found in Table 2, is as follows:

Table 2. Risk Values

	Assessment of Equity	*Management of Power*	*Communication of/for Power*
Scientific Practice	Lowest Risk for the Greatest Number	Administrative Decisionmaking	Information Transfer
Social Theory	Greatest Good for the Greatest Number	Consent to Power or Trust in Power	Deliberative Participation

Although this project is mainly descriptive, there is an important prescriptive element to it as well. That is, the foundational liberal theories of risk should be understood in order to rerationalize the practice of risk. However, I am not proposing that liberal and rational divisions of risk should be collapsed completely into one another because they perform different, yet necessary, functions. Rather, in order to combat the current undemocratic character of environmental policy, each group needs to understand and consider the opposing group's position. If scientific practitioners of risk clarify the values of their bureaucratic agencies, American society might be better able to interpret and react appropriately to the risks it actually faces with respect to food biotechnology. Additionally, defining the liberal objectives of the American public will allow bureaucracies to appreciate the reasonable concerns of society and the appropriate limits of its practices.

CHAPTER TWO

THE TECHNICAL VALUES IN THE EXPERT JUDGMENT OF RISK

1.INTRODUCTION

At the first United States' biotechnology conference in 1975, molecular biologists unanimously consented to follow regulations established by the Recombinant DNA Advisory Committee (RAC) of the National Institutes of Health (NIH) to review, monitor, and approve all research projects in the brave new field of genetic manipulation. Notwithstanding the incredible economic potential of biotechnology for pharmaceutical and agricultural companies, the RAC recommended that industry experts refrain from some experiments and follow stern restrictions for others. For example, in 1984, when the advisory committee introduced "non-scientific and non-objective considerations into the guidelines about genetic transfer experiments," it declared that research involving the transfer of genetic traits between animal and human germ lines to be "morally and ethically unacceptable" (Naik, 2000, 38). In 1986, at the suggestion of the RAC, the United States government created a coordinated framework of governmental agencies and industry experts to ensure that biotechnology-derived food products are substantially familiar or equivalent to their non-biotech counterparts. By maximizing many of the political benefits while minimizing most of the societal risks, the RAC successfully monitored the ethical impacts of genetically modified organism research and development (Naik, 2000, 59). According to some risk scholars, strict governmental scrutiny of biotechnology research, corresponding deliberate industry development, and the transparent public actions of the RAC represent a now lamented past that is conspicuously different from the present (Krimsky, 1991, 183-184).

Before any genetically modified food products were approved, executives of the leading agricultural biotechnology company, Monsanto, met with Vice President George Bush and asked the Reagan administration not to follow its policy of governmental deregulation. What they wanted instead was highly visible federal oversight of the new science of biotechnology (Eichenwald et al., 2001, 2). Aware of environmental concerns, ethical challenges, and consumer questions, Monsanto executives also proposed to "encourage public interest" by having consumer and environmental groups serve as formal advisors and consultants to Monsanto in order to help "develop supportive positions on biotechnology" (Eichenwald et al., 2001, 6). Industry experts reasoned that the public would have confidence in the products and processes of the most radical industry since nuclear power if biotech researchers consulted adversaries, were controlled by the RAC, and supported the regulations enforced by the USDA, EPA, and FDA.

Shortly after articulating this progressive outreach strategy of having political opponents advise the biotech industry and asking for strict regulatory boundaries, Monsanto decided to introduce its first commercial genetically modified food organism, rBST. "Despite vociferous objections of executives who feared that tinkering with a product consumed [mainly] by children would ignite a national outcry," the company introduced the United States to biotech food with a product that benefited certain farmers, not consumers (Eichenwald et al., 2001, 7). As a consequence of the enormous controversy surrounding the recombinant bovine growth hormone, both in America and abroad, Monsanto abandoned its commitment to policy inclusion with public persuasion and introduced a strategy of policy exclusion and expert influence—a decision that many company executives now describe as a mistake (Shapiro, 2000).[3]

Under the advice of a new management team at Monsanto, the government "ushered through an unusually generous policy of self-policing" for the contested technology industries and decided to loosen the power of federal regulatory agencies in order to speed biotech foodstuffs to market (Eichenwald et al., 2001, 2). The 1991 governmental "Report on National Biotechnology Policy" by the President's Council on Competitiveness, reads as follows:

> In biotechnology, as in many other high technology industries, Federal regulation is a critical determinant of the time and cost to bring a product to market. In serving as "gatekeepers" for the development and use of new products, regulatory agencies may create substantial barriers to product development. These barriers result from the costs of testing to meet regulatory requirements, the potential for delay in regulatory approval, and the uncertainty associated with the possible imposition of extensive restrictions or outright disapproval of new biotechnology research or products. In addition, uncertainty related to the extent of effectiveness of Federal regulation may lead to the enactment of a patchwork of conflicting and burdensome state regulations. Delay, cost, and regulatory uncertainty discourage new research in regulated areas and curtail the development of new products, as well as undermine public confidence. In

[3] Some scholars would argue that public opposition to rBST was also caused by other events. For instance, many would reference the volatile reaction to Advanced Genetic Sciences strawberry and potato plants that were bioengineered with the "ice-minus" bacteria. Engineered to resist frost, test plants were awkwardly monitored by scientists in full body spacesuits equipped with oxygen masks and uneasily regulated by the EPA who declared oversight only by naming frost a pesticide.

general, to avoid unnecessary burdens on biotechnology, the Administration has sought to eliminate unneeded regulatory burdens for all phases of developing new biotechnology products—laboratory and field experiments, products development, and eventual sale and use. Existing regulatory structures for plants, animals, pharmaceuticals, chemicals and toxic substances provide an adequate framework for regulation of biotechnology in those instances where private markets fail to provide adequate incentives to avoid unreasonable risks to health and the environment. In these instances, regulation also can help shield industry from avoidable incidents that could tarnish its image and impair its development (Quayle, 1991, 11).

The administration's policies of limited risk-based regulation seemed to value biotechnology companies by not inconveniencing them with strict rules that might protect society from uncertain environmental or health risks from the new industry. Furthermore, this approach to governmental oversight represented a significant transformation in risk policy because it treated agriculture issues as economic-scientific decisions that could be made with objective certainty rather than liberal-social decisions with normative or subjective concerns.

Then in 1992, under the Bush Administration's "Regulatory Relief Program," Vice President Dan Quayle announced that the United States "will speed up and simplify the process of bringing better agricultural products, developed through biotech" by allowing them the "same oversight as other [conventionally derived agricultural food] products, instead of being hampered by unnecessary regulation" (Eichenwald et al., 2001, 9). The new policy limited the regulatory scope of the FDA by asserting that biotech foods would be tested only if the companies producing the new products conducted the tests and, further, that any genetic labelling by the government or the food industry would be considered illegal because it would mislead the consumer about unintended effects and unscientific risks of universally safe products and processes of food biotechnology (FDA, 1998). Monsanto, Eli Lilly, Upjohn, Aventis, Syngenta and other huge biotech companies who had spent billions of dollars to create genetically altered organisms used in modern agriculture argued that their industry testing was superior to anything the government could conduct because of the incredible operating costs involved in doing risk assessments (Barboza, 2001). Furthermore, they argued that a self-testing approach is more cautious than a full government regulatory review because biotechnology industries have a lot to lose if their science is incorrect.

That 1992 policy has essentially remained unchanged. For instance, although the Clinton Administration required governmental review of some previously confidential industry tests, this reform was largely symbolic because companies had always submitted such data under the voluntary rule. Similarly, no risk practices were affected in 2001 when the Bush Administration revised industry guidelines on food labels by making producers use common terms in order to help consumers who are highly allergic to some food ingredients (Winter, 2001). The new food labels, however, are not mandated by the FDA and do not include the identification of any genetically altered organisms that may cause adverse human health reactions or unintended environmental damage.

The only consistency in this brief history of American bureaucracy and biotechnology across four presidential administrations is that government complied with industry desires. When the burgeoning genetic engineering companies demanded regulations, they got the RAC; when they requested industry freedom they were rewarded with the 1992 framework of Regulatory Relief. One result of the coordinated industry-government support of biotechnology, however, has been incredible national economic success. It has been said, for instance, that the "social and commercial effects of transgenic species" will be the "raw resource of the new economic epoch" in American history (Beaudoin, 1999, 242). Genetically modified organisms are in about half of America's groceries and more engineered products and processes are under patent review than have already been approved. United States' farmers are projected to increase the amount of modified seeds they plant every year for the next decade, and agribusinesses are reporting some of the largest fiscal gains in their thirty year history. Agricultural experts also contend that large countries with policies that limit or ban the consumption of genetically modified food products may soon have to change course, because they will not be able to grow or import enough non-biotech food to meet their needs (Barboza, 2001).

Cognizant that several European countries have rejected food biotechnology, that many American companies have voluntary rejected altered agriculture, and that consumer and environmental protestors have expressed moral opposition to changing nature, the biotechnology industry has started a multi-million dollar advertising campaign that explains the potential to feed the growing human populations around the world with their revolutionary farming practices and products. Industry spokespeople for Monsanto, furthermore, have announced a new policy strategy of consulting with groups concerned about biotechnology. As a result, Monsanto's 2000 annual report for stockholders includes the following pledge:

> We will respect the religious, cultural and ethical concerns of people throughout the world by not using genes from animals or humans in products intended for food or feed. We will never commercialize a product in which a known allergen has been introduced. We affirm our commitment not to pursue technologies that result in sterile seeds. We will use alternatives to antibiotic-resistant genes to select for new traits as soon as the technology allows us to do so. We will commercialize commodity grain products only after they have earned approval for consumption by both humans and animals (Monsanto, 2002).

This outreach program of "dialogue, transparency, respect, sharing, and benefits" is suspiciously close to their corporate strategy that was abandoned with the introduction of rBST (Monsanto, 2002). More important, this market strategy is significantly dissimilar to the arrogant approach taken by biotechnology companies during much of the 1990s. A Monsanto advertisement poster that shows a cow eating a five-dollar bill with the quote, "Subtle Reminder of What You're Losing On Each Cow That's Not Being Given Posilac," is possibly the best illustration of the multi-billion dollar industry that failed to consider the social consequences of modern agriculture (Steyer, 1996). All of these strategies are important to analyse, however, because they offer a window into the dominant rational thinking the technical risk community uses to administer development of the genetically altered food industry. Making the theoretical foundations of the technical elites explicit is important because it may encourage technical experts to be self-critical of their risk work and aid public understanding about how technical assessors rationally think about risk.

The lack of analysis of rational risk management in the agricultural-biotechnology industry is best seen in the new television campaigns that explain all of the potential for food biotechnology, including possible pharmaceutical benefits to children with birth defects, cancer patients, hepatitis B sufferers, and elderly diabetics. Their implicit argument seems to be: "Diabetics don't care if their insulin is genetically engineered, so why should the public care if their food is" (Martineau, 2001, 24)? This line of reasoning, however, fails to grasp the rudimentary understanding that: "sick people are willing to take a much higher risk with drugs to regain their health than healthy people are willing to take with their food"(Martineau, 2001, 24). Industry experts who believe that the flawed insulin example will motivate wary consumers to trust the powerful biotech industry fail to understand seriously the values that underlie public assessments of risk.

This section of the book will describe the supporting philosophical arguments or value reasoning behind the policy positions of biotech companies and many of their allies—numerous big farmers, most American universities, many food companies, and almost all of the regulatory agencies of the Federal government. From the viewpoint of rational or scientific experts I will carefully consider the value judgments made about what the issues are, who benefits, and how much is at risk in modern agrarianism. Proponents of new biotechnological methods in agriculture will discuss the economic and social effects of biotechnology on agriculture. The acceleration toward fewer and larger farms with much specialization will also be discussed. Additionally, the agribusiness' stances on political issues like governmental regulatory testing, food labelling, research ethics, and patenting of biological products will be analysed.

The theory advanced herein is that in order to manage complex decisions beset with uncertainty, industry and bureaucratic experts adopt the policy strategy of using scientific rationalism in risk assessment. It is important to note that expertise is a resource that gives political legitimacy to technical assessments because the public incorrectly assumes that all expert decisions are made with scientific certainty. This book does not discredit the scientific data used by industry and government risk assessors involved in the safety tests of food biotechnology. Rather it contends that some expert decisionmaking that is based on definite science has significant political consequences that are not considered. For instance, scientific experts who assert that new food biotechnology techniques are extensions of old plant and animal breeding practices are technically correct. Nonetheless, because the new methods of biological sciences are much more focused on food production than the old techniques of breeding, they have more economic, social, and political consequences to be analysed (Jackson & Piper, 1989, 1592). That is, risk assessors who concluded that rBST would safely increase milk production were not incorrect, but rather were woefully incomplete because rBST: (1) increased profits in dairy farming while decreasing the profitability of milk production; (2) established the standing legal practices for commercial sales and marketing of biotechnology products; and (3) created the regulatory structure for all genetically modified foods.

2.THE ECONOMIC EFFECTS OF RBST USE

In the almost eight years that rBST has been used by the American dairy industry, there has been no reported case of human sickness due to the consumption of milk or meat from cows treated with it (FDA, 1999a). After investigating the 806 "adverse-reaction" reports from thirteen thousand farmers, the FDA also concluded that there was little to no unusual animal effects or problems associated with rBST (Powell & Leiss, 1997). Similarly, tests conducted by Monsanto and reviewed by the FDA on dairy "herds over 24 months showed no difference in health problems between BST-treated cows and untreated cows" (Steyer, 1997). More importantly, the predicted decline in milk consumption, increase in antibiotic residues in milk, and willingness of the majority of American consumers to pay more for rBST-free milk has not happened.[4] As an alternative to all of the negative consequences predicted by

[4] This sentence is factual in that all of the predictions of rBST opponents were incorrect. That is, consumers have not abandoned buying dairy products because of the bovine growth hormone. It should, for instance, be noted that the average American drank 225 pounds of milk in 1996, which is

critics, "rBST has emerged as a niche drug used on certain animals during certain stages of lactation. This dairy farming management tool produces the same amount of milk with fewer cows, yielding a small but incremental environmental benefit and economic incentive for dairy farmers"(Powell, 1999).

Despite the controversy, rBST appears to be just another technological advance that the dairy industry has implemented in the past half century, according to market analysts. They note that since 1950, technological innovations like recombinant bovine growth hormone, "artificial insemination, new feed additives, mechanical milk-handling equipment and computerized tracking of production" have allowed annual milk production "to increase 33 percent with 57 percent less cows" (Steyer, 1996). Assertions that rBST would result in a huge increase in milk production that would cause dairy prices to plunge and surpluses to soar were erroneous, and some argue that economic changes since its introduction are more the result of geography, feed costs, forage problems, weather, and types of dairy products on the market than any direct impact of rBST (Butler & Cohn, 1993). A huge surplus, followed by a price collapse, could only happen if Congress changed federal subsidy laws and state legislatures altered financial assistance programs in order to make farmers more subject to the marketplace.[5]

The most negative social consequence of rBST, according to expert and lay observers alike, is that the newly applied agricultural technology produces fewer and larger dairy farms. Metaphorically speaking, it was feared that rBST would create a treadmill market wherein the small American dairy farmer could no longer run with larger dairy businesses. While Robert Kalter predicted social disruption in rural America by the specific process of high technology dairy farming (Kalter, 1985; P. Thompson, 1998b), the more descriptive and general economic theory of the "technological treadmill" by Willard Cochrane is as follows:

> The aggressive, innovative farmer is on a treadmill with regard to the adoption of new and improved technologies on his farm. As he rushes to adopt a new and improved technology when it first becomes available, he at first reaps a gain. But, as others after him run to adopt the technology, the treadmill speeds up and grinds out an increased supply of the product. The increased supply of the product drives the price of the product down to where the early adopter and all his fellow adopters are back in a no-profit situation. Farm technological advance in a free market situation forces the

---

down two percent since Monsanto's Posilac has been on the market. However, the declining fluid milk consumption has been falling for decades, for example it is down 31 percent from 324 pounds in 1960. According to dairy economists at the USDA, this is due to the demographical shift of fewer younger children per year, the largest cohort of milk consumption in the United States.

[5] Any immediate change in federal law is doubtful. The basic formula for liquid milk prices and surpluses has remained significantly unchanged and the minimum market price for manufactured milk--used in butter, cheese, ice cream and other products—is well above the government support price despite pulling back on the federal safety net in both 1985 and 1990.

> participants to run on a treadmill...Not only did the process of farm technological advance force the participants in the process onto a treadmill, but it created a condition in which the strong and aggressive farmers gobbled up the weak and inefficient. The process of farm technological advance has contributed importantly to the redistribution of productive assets in American agriculture in which commercial production has been, and continues to be, concentrated on the larger farms (Browne, Skees, Louis, Thompson, & Unnevehr, 1992, 56).

The agricultural economics of food production resemble many other capitalist enterprises. That is, in order to remain competitive, one must continually adopt new industry technologies, not just for short-term gains, but also to avert the long-term risk of operating at a comparative disadvantage to others in the same field who have already adopted the beneficial technology.

Despite the fact that the United States is the world's largest agricultural supplier, the 2000 census reveals that for the first time in American history farming was not considered a recorded occupation because the agriculture population has been reduced to less than two percent. The transformation of small family farms to large corporate agribusinesses is a well-studied sociological phenomenon that began to happen between 1970 and 1980, long before the commercial release of biotechnological food products. Therefore, while genetic engineering can be implicated in the continuing trend toward agribusiness consolidation, it cannot be the causal link. Furthermore, despite projections that rBST has the potential to reduce the capital investment, operating costs, and profitability of milk production, it is clear that rBST alone has not created the phenomenon of fewer and larger dairy farms. Monsanto claims that rBST is the largest selling dairy animal health product among both large and small dairy producers. This includes farmers who have herds ranging from 5 to as many as 10,000 cows (Monsanto, 2000). In fact, Monsanto claims that 65 percent of rBST customers have less than 150 cows (Monsanto, 2000).

Although numerous social scientists have attempted to measure the social impact of rBST on the economics of milk production, none have convincingly proven that small farms have gone out of business because they failed to adopt the new and controversial recombinant bovine growth hormone drug (Tauer, 1992). In fact, scientific studies have failed to prove that rBST is economically profitable for dairy farming. For instance, although data from a panel of New York dairy farms found that rBST significantly increased milk output per cow, the impact on profits for the operating farm were not substantially positive in the aggregate effect (Stefandies & Tauer, 1999, 95). That is, the use of rBST was not profitable, on average, for these farms because the drug is either not a strong enough explanatory variable for influencing profit or there is a longer business learning phase needed with administering the new technology.[6] Although genetic engineering generally and rBST specifically may be influencing the treadmill of agricultural technology, it does not appear to be the most important advance implicated in changing the agricultural countryside of America's dairyland.

Precisely measuring the economic impact of rBST will take much longer than anyone—backers, critics, and scholars—ever expected. Part of the difficulty in assessing the impact of rBST is due to the incomplete information available to agricultural economists and political scientists. For instance, no definitive answers can be found to questions about how many American farmers and cows use Posilac, Monsanto's trade name for its recombinant bovine growth hormone, or what percentage of American milk is produced with rBST technology. Since rBST milk is virtually indistinguishable from non-rBST milk and because most dairy farms pool their milk, the latter question is impossible to answer. With regard to the former, only the following can be discerned: "Monsanto will only say that at least 17,000 dairy farmers, or more than 15 percent of U.S. dairy producers, have used the drug since February 1994. But that's a cumulative figure. Monsanto doesn't say how many farmers tried BST and then quit. Monsanto also doesn't provide estimates on the number of BST-treated cows" (Steyer, 1997). The Food and Agricultural Policy Research Institute says 20 to 25 percent of U.S. dairy cows were treated with BST in 1998, but projected figures from the USDA say 13.25 percent of U.S. cows or 1.28 million cows are injected with the bovine growth hormone (Steyer, 1997). More recently, however, Monsanto claimed the cumulative figures of 30 percent or three million cows of the U.S. dairy herd (Monsanto, 2001b).

In correspondence with not offering any more information than it is legally obligated, Monsanto will not disclose any revenue information on the specific sales of rBST. It is known, however, that the average cow requires 18 shots of the veterinary drug at a list price of $5.80 a dose. Considering that discounts are offered if farmers commit to using the genetically modified technology on a significant percentage of their herd for a half of a year, it is estimated that 23 million unit sales were made in 1997, producing an estimated $120 million in revenue and permitting Monsanto to make a profit on sales of Posilac for the first time (Steyer, 1997). The latest public information released from Monsanto claims a 25 percent increase in sales in 1998 and 20 percent in 1999, and boasts of selling the 150 millionth dose (Monsanto, 2001b). Regardless of these numbers, farm policy experts argue that rBST is not even close to recouping the economic investment Monsanto made in research and development beginning in the early 1980s. Industry analysts, however, argue that the ultimate value of rBST will be in the

[6] A similar study is found in the December 1999 Journal of Dairy Science.

lessons Monsanto can apply to its other biotechnology products, sales, and businesses (Steyer, 1997). As seen in subsequent chapters, unlike social opponents of food biotechnology and governmental regulators, Monsanto seems to be the only policy stakeholder that has learned anything from the rBST case.

3.COMMERCIALISATION OF THE BIOLOGICAL SCIENCES

The original pursuit of Western science has undergone two dominant paradigm shifts in recorded civilization (Krimsky, 1991, 1999). The first occurred during the "seventeenth and eighteenth centuries when the distinction was made between 'experiments of light' that seek to discover the cause of things, and 'experiments of fruit' that apply the knowledge to practical ends'" of things (Krimsky, 1999, 2). The second came in the years immediately following World War II when the transformation of science to its current objective as an economic pursuit occurred—facilitated by academia, encouraged by public policy, and aided by industry funding. Today, it is clear that the pre-nineteenth century vision of science as a community of professors discovering how things work has been abandoned for a view of science as technicians researching how to develop things that work. (Crosby, 1995; Jackson & Piper, 1989). Nowhere is this shift more apparent than in the environmental sciences and the industries of biotechnology.

Actually, biotechnology—or life technology—is by its very definition applied life science, and since its genesis it has been a combined effort of scientific discovery and technological innovation. More importantly, since Herbert Boyer helped develop the recombinant DNA technique at the University of California, and then started the first genetic engineering company Genentech, and patented the biological process of rDNA, biotechnology has been the amalgamation of not only science and business, but also politics (Krimsky, 1999). Economists argue that American research into and development of biotechnology has been able to expand continually beyond most business model expectations because it is mostly a collaborative effort by industries, universities, and governmental agencies (Falcon, 2001). The fruitful marriage of corporations and colleges is by no means new or immediately ethically bothersome, for most researchers.[7] If truth were told, most university professors concede that molecular biology

[7] Jeremy Rifkin, Michael Fox, Martin Titel, Ralph Nader and other environmental activist-scholars discussed in chapter five would be offended by the notion that environmental science needs corporate financing and that their marriage is a moral union. Although the relationship of biased research and corporate funding is a topic worthy of significant concern and investigation, the collaboration of business and environmental science is a necessary and beneficial union.

research would be impossible in academia without corporate grants, and that company scientists conduct the best research in biotechnology because they have the greatest amount of resources—money, time, and freedom—for discovery.

There is an emerging ethical concern, however, regarding government's role in promoting the commercialisation of the biological sciences. Although anecdotal, the ethical problem of governmental science is nicely illustrated by the recent "Cooperative Research and Development Agreement". It allows government risk assessors to conduct industry tests in federal laboratories and receive royalty income for product ideas, and to protect their trade confidentiality, the agreement exempts the assessors from government requirements to publicly document their findings (Krimsky, 1999).

While the legislative support for biotechnology is too unspecified to be descriptive, the governmental support of the scientific-commercial field of applied genetics can be analysed by examining a series of legal battles and court cases. Supplementing the federal agencies' coordinated framework for the regulatory relief of biotechnology in 1992 is the United States court system's legal doctrine that seeks not to impede new technology without cause. Thus, government's supportive role in the development of biotechnology is apparent in the legal rulings that redefine patent law, circumvent environmental requirements, limit information on product labels, and establish intellectual property rights beyond traditional confines.

The first and probably most significant case is the 1980 Supreme Court decision of *Diamond v. Chakrabarty.* Article 1, Section 8 of the United States Constitution grants patents to inventors "to promote the Progress of Science and useful Arts," and the utility patent statute 35 U.S.C. 101 provides a patent protection for a seventeen year period for "whomever invents or discovers any new and useful process, machine, manufacture, or composition of matter, or any new and useful improvement thereof"(Ewens, 2000, 292). Prior to the groundbreaking Chakrabarty case, however, the U.S. Patent and Trademark Office (PTO) did not allow living matter to be granted a utility patent.

In a 5-4 vote, a divided U.S. Supreme Court ruled that a new bacterium created by the microbiologist Chakrabarty and General Electric to break down crude oil for easier cleaning after an inadvertent spill "was a product of human labor and contained characteristics markedly different than any found in nature, and [therefore] showed the potential for significant utility"(Ewens, 2000, 294). By making a distinction between products of nature and products of human effort, the court opened the door for biotech companies to protect scientific discoveries for commercial progress (Goss,

1996, 1405). With Chakrabarty and its related casework, biotech companies have been able to define new limits to existing intellectual property rights.

Through the use of recombinant DNA technology, University of California and Advanced Genetic Sciences, Inc.(AGS) researchers in the early 1980s modified a bacterium to delete the ice-making trait of an organism in order to induce frost tolerance. If the genetically modified bacterium was inserted into a plant seed it could prevent certain agricultural crops from freezing during a pre-harvest frost. The "ice-minus bacteria" research was initially stopped by plaintiffs who sought an injunction on the grounds that (1) the NIH illegally approved the university field experiments which would result in the deliberate release of genetically engineered organisms in violation of the National Environmental Protection Act (NEPA) that required an Environmental Impact Statement (EIS) for such a study, and (2) the EPA "arbitrarily and capriciously granted a permit" for AGS that violated the Federal Insecticide, Fungicide and Rodenticide Act's (FIFRA) requirement for a detailed risk assessment of any naturally unknown substance (Naik, 2000, 70).

In both the Federal District case of *Foundation on Economic Trends v. Heckler* and the Appellate Court decision in *Foundation of Economic Trends v. Thomas,* it was decided field studies could be conducted contingent upon minor revisions and that,

> "the use of delaying tactics by those who fear and oppose scientific progress [was] nothing new." Furthermore the judge worried that a "national catastrophe" would result should the development of the promising technology of genetic engineering be "crippled by the conscionable delays. . . brought about by litigants using [environmental regulations].(Naik, 2000, 74)"

The advent of biotechnology has revolutionized American society in many ways; however, it appears that occasionally the promises it offers for the future has led the legal system to excuse some present environmental concerns.

Recognizing consumer petitions and a lack of federal guidance on label disclosure of genetically modified organisms, "the Vermont legislature enacted a BGH labeling scheme, which involved positing a BGH-information sign in retail outlets accompanied by a blue rectangle affixed to BGH-produced products"(Beaudoin, 1999, 254). This went beyond the interim guidelines issued by the FDA, which state that voluntary labels must be truthful, not misleading, and contains a qualifier that the FDA has approved the drug. However, the Second Circuit Court of Appeals in *International Dairy Foods Association (IDFA) v. Amestoy* held that Vermont's labelling was not voluntary, did not have a substantial state interest to bias the fair bargaining process of commercial dairy, and could confuse the American citizenry on the FDA's position regarding the safety of rBST.

Unless information is deemed to aid in the health, safety, or substantial knowledge of the consumer, the standard for biotech labelling appears to favour technological innovation and economic advantage over the concerns of its democratic citizenry. Despite an overwhelming number of Vermont consumers wanted recombinant DNA information on their food products and regardless of much IDFA evidence questioned the safety of rBST the Amestoy court found that dairy producers have a first amendment right of commercial speech to not disclose certain information about their product. Claiming that protecting the welfare of consumers against unknown risks of genetically engineered food is an illegitimate state interest, the Amestoy court placed a strict standard on disclosure requirements for GMO products.

In two similar cases, *Cordes v. Madigan* and *Barnes v. Shalala*, dairy farmers and consumers from Wisconsin accused the FDA, USDA, and the National Dairy Promotion and Research Board of encouraging the use of rBST without: (1) examining all the economic, health, and safety consequences of the new drug; (2) requiring mandatory labelling on products released from rBST treated cows; and (3) including a long-term EIS in governmental risk assessment documents (Naik, 2000, 108). In a pair of confusing judgments, the Wisconsin court ruled that with the exception of consumers and the issue of labelling, the plaintiffs failed to have a concrete and particularized standing and "while it was ready to discuss health, safety and environmental concerns, it was not willing to entertain other concerns about the nature or mistrust of corporate power and government, which were among the main concerns motivating" the suit (Naik, 2000, 110).

The obvious continuation of the aforementioned cases culminated in the *Stauber v. Shalala* decision. Wisconsin citizens challenged the FDA's decision not to require labelling of rBGH-treated milk under the Food Drug and Cosmetic Act (FDCA) of 1938, an Act which ensures that consumers are informed in their purchase decisions (Naik, 2000). The Wisconsin residents in this suit contended that if the fundamental purpose of a label is to inform or instruct consumers, dairy products should have food labels that identify them as wholesome or engineered. They contended that such labels would provide valuable information to some morally and politically conscious consumers who may choose not to purchase rBST products because of the attendant social and economic consequences on modern dairy farming. Following legal precedent and the federal policy of biotech advocacy, the District Court of Western Wisconsin observed:

> If there is a difference, and consumers would likely want to know about the difference, then labeling is appropriate. If, however, the product does not differ in any significant way from what it purports to be, then it would be misbranding to label the product as different. In the absence of evidence of a material difference between rBST-derived milk and ordinary milk, the use of consumer demand as the rationale for labeling would violate the [regulatory guidelines of the FDA] (Beaudoin, 1999, 262).

In reviewing the evidence regarding conflicts between consumer rights and state interests in economic or social effects of a new technology, the court system seems to defer to the FDA and consider only so-called scientific grounds for legal reasoning about labels.

In a continuing and perhaps final attempt to require labelling of genetically engineered organisms, dairy manufacturers in many states began to produce their own voluntary labels. While most companies followed the nondescript guidelines for voluntary labels, Ben and Jerry's, Stonyfield Farm, Whole Foods Market, and Organic Valley sold products under an "rBGH free" label. They were taken to court by Illinois officials who claimed that the practice violated the FDCA, which forbids packaging information that may mislead the consumer (Beaudoin, 1999, 267). Ben and Jerry's and the three other organic food companies were forced to compromise the language of their statement so as not to imply that milk from rBST treated cows is somehow tainted (Reporter, 1997). An example of how the revised label now reads follows:

> WE OPPOSE RECOMBINANT BOVINE GROWTH HORMONE. The family farmers who supply our milk and cream pledge not to treat their cows with rBGH. The FDA has said no significant difference has been shown and no test can now distinguish between milk from rBGH treated and untreated cows. Not all the suppliers of our other ingredients can promise that the milk they use comes from untreated cows.[8]

Anti-biotechnology, environmental, and consumer groups continue to petition the FDA to remove the FDA qualifier and question if any governmental agency has the ability to force a food manufacturer to package its products in such a biotechnology friendly manner.

With the notable exception of the new organic labelling system approved by the USDA, the United States federal court system remains coordinated with the executive and legislative branches' pro-biotech positions.[9] That is, in order to sell certified organic milk in the U.S. a dairy farmer must not treat his or her cows with hormones and antibiotics or feed them grain and hay with herbicides, pesticides, or chemical fertilizers. Retail sales of organic dairy products are growing about twenty percent per year and nationally the number of dairy cows has jumped from 13,000 in 1997 to 49,000 in 2001 (OCA, 2003). However, very few Americans purchase organic dairy products because a gallon of conventional milk is typically half the price of a gallon of organic milk. In order to be both profitable and progressive some

[8] Label is found on All Natural Ben and Jerry's Vermont's Finest Cherry Garcia Frozen Yogurt product dated 1999

[9] The USDA's first proposed definition for organic food included irradiated, genetically engineered, or grown with sewage sludge fertilizer. According to some environmentalists it was because the USDA received over 200,000 complaints that its final definition did not include irradiated, engineered, or sewage fertilized food.

inventive dairies—like Oberweis and Oakhurst—have been successful at selling non-organic dairy with the pledge that their milk does not come from cows injected with rBST.[10] Dairy products marketed as such, however, must follow case law and use the aforementioned rBGH free label with disclaimer.

In the legal realm, the last prominent development affecting biotechnology products and processes is found in the promulgation of the first significant case of genetic engineering, *Diamond v. Chakrabarty*. Because Chakrabarty dealt with a bacterium, it did not secure the use of utility patents for plant breeding innovations, economically the largest area of the biotech industry (Ewens, 2000). The seed breeding market is dominated by large chemical producers, and a perfect example is Monsanto, the second largest seed company in the world (Pistorius & Wifk, 1999). Annually it spends over six billion dollars on the life sciences industry—crop research, development, litigation, and marketing (Ewens, 2000, 290). Although Congress had enacted the 1930 Plant Patent Act (PPA) and the Plant Variety Protection Act (PVPA) of 1970 to provide intellectual property rights to plant breeders, the protection they provide is not as secure as a utility patent because of the research and crop exceptions allowed farmers (Pistorius & Wifk, 1999). A legal scholar explains:

> Unlike the utility patent statute, which allows the patentee to exclude anyone from making, using, or selling her invention, the PVPA allows a breeder's competitors to use her protected variety or create new varieties without their permission. Furthermore, the PVPA allows farmers to save seed from crops grown from a breeder's protected variety and use the seed for certain purposes without compensating her (Goss, 1996, 1406).

With the aid of some important litigation and legislation, however, the scope of the two exceptions is being narrowed to protect many biotechnology companies' innovations.

For instance, the crop exception allows farmers to save seed grown with a protected variety to replant it in the following year or sell it to another farmer for replanting in order to develop or research their own crop varieties. While this loophole seems innocuous, Dennis and Becky Winterboer of Iowa sold 10,000 bushels of the PVPA protected Asgrow soybean seed, enough to plant 400,000 acres, with no attempt to create their own research lines or compensate the Asgrow seed company (Goss, 1996, 1412). The loophole was closed to some extent, however, by the courts. In the 1994 *Asgrow v. Winterboer* case, the Supreme Court ruled that because the Iowa farmers were growing the patent-like protected soybeans exclusively to sell them, and not to develop their own research, they had infringed upon the PVPA

[10] Please see http://www.organicconsumers.org/rbgh/rbghlist.cfm/ for a list of organic and non-organic rBST-free producers in the United States.

certificate. While the crop exception protects the traditional practice of saving seed by farmers, it contradicts the intent of the PVPA, which is to encourage the research of new plant varieties.

The lessons of Winterboer, however, are more significant than extending the intellectual property rights of plant breeders; it has encouraged Congress to pass amendments to the PPA and PVPA that narrow the gap of the research and crop exceptions. Additionally, it has inspired biotech companies to render seeds sterile after one use, with so-called "terminator technology," or insert marker genes that identify plants' DNA strains as the intellectual property of a particular biotech company. The recently patented terminator seeds biologically guarantee what has been legally impossible to enforce through property rights, the absolute monopoly on a seed's biotech code by destroying its reproductive capacity (Ewens, 2000, 307). While the legality of both the amendments and the terminator traits of biotech seeds are currently under review in the U.S. court system, the outcome seems predictable. United States federal policy, agency regulations, and case history all value and protect innovation by offering incentives for investment in research, financial development, and social progress.

The commercialisation of the biological sciences raises a myriad of questions including: (1) ethically, should any life—plant, animal, or bacteria—be patented? (2) legally, how can the biotechnological arranging of old genes among old organisms be new? (3) economically, is it possible to find equilibrium between farmers who grow crops and multinational corporations that hold the patents to those same crops? (4) environmentally, will an increase in the use of organism varieties contribute to the decline of genetic biodiversity (Ewens, 2000)? The easy political answer to all of these ethical, legal, economic, and environmental questions can be found in the philosophy of utilitarianism or the rationale of consequentialism. That is, even if the above questions have some merit and "biopatents are disrespectful of life, misconstrue the nature of biotechnical innovation, and render inappropriate deserts for such innovation, the tremendous social utility of biopatents outweighs these considerations" (Hettinger, 1995, 291). In policy debates over the above questions, therefore, one must simply state: If Americans want oil-eating bacteria, cheaper milk, and herbicide resistant soybeans—not to mention wonder drugs, food products with less saturated fats, and high yield crops that could eliminate world hunger—we must allow scientists to patent these innovations and sell them in the market (Hettinger, 1995, 292). Put another way, without governmental protection, copycat competitors would undersell innovator companies. Knowing this, companies would back off from innovation because recouping biotech product research and development costs would be less likely, and the American society would never be able to harvest the potential economic, scientific, and social benefits of biotechnology.

4.POLITICAL JUSTIFICATIONS FOR FOOD BIOTECHNOLOGY

Proponents of biotechnology contend that the intellectual radiance of the new biological sciences is not found in the products of genetically modified organisms, but in the process of genomics—the classification, location, and function of all of the genes for a species. In addition to being a major breakthrough in the science of classical breeding for agriculture, it is offering enormous potential for human disease research and is responsible for removing the fiction from the science-fiction narratives of genetic sequencing and cloning. Despite the controversies surrounding all of these advancements, an overwhelming number of people and institutions maintain that the science of genomics plays a major role in solving the tragic drama of the world food supply. According to many convinced stakeholders and policymakers, although biotechnology has some known and unknown risks, as any new science and technology does, they can and should be managed to solve many of the earth's natural resource problems.

According to many scientific experts, educating the public about the science of genomics may bridge the gap between the rational understanding of risk by technicians and the irrational perception of risk by the public. For instance, McHughen's book, *Pandora's Picnic Basket: The Potential and Hazards of Genetically Modified Foods,* explains that all living organisms have the same DNA. The differences, however, are found in the number and arrangement of the genes in the cells that make up the living creatures. People, therefore, should not be alarmed by the transferring of like genes between unlike organisms because this is just a quicker and safer method of agricultural crossbreeding, which has been done for tens of thousands of years. Indeed many argue that there are more scientific risks in the traditional crossbreeding of species than the gene splicing of modern agriculture. Furthermore, some experts contend that the next several decades of transgenic modifications will not alter living organisms as significantly as centuries of traditional breeding have (Hettinger, 1995).

The ancient methods of crossing two parents with different desirable traits in order to produce a new and better offspring have given farmers such products as dwarfed grain, sweet corn, and seedless oranges. Similarly, strawberries grown in the northern United States that have the antifreeze gene from the Arctic flounder will be given an extended growing season, not some fish-like traits. University plant breeders suggest that the public understanding of the enormous potential of food biotechnology could advance if people thought of genes in terms of their function and not their origin (Brody, 2000). Accordingly, they would argue that it is not a fish gene but a cold tolerance gene that was introduced into the new variety of strawberries (Brody, 2000). Experts contend that public confidence will increase when the consumer understands the food benefits of genetic engineering. Whereas future products like foods that deliver vaccines or contain fewer toxicants may booster consumer confidence, currently the promise of food biotechnology to solve the problem of world hunger is gaining the most amount of public trust.

The list of biotechnology supporters and promoters united around the issue of the world's food supply is an amazingly impressive group, albeit a motley crew. For instance, the United Nations (U.N.) has publicly credited the multinational corporations involved in agricultural biotechnology with helping to give real answers to the world's abstract health and hunger problems. The U.N. Food and Agriculture Organization maintains that the ability of genetic modification to make agricultural crops and livestock stronger, higher yielding, more resistant to disease and insects, and more nutritional are important to the developed, developing, and underdeveloped countries of the world.[11] The U.N. Development Program supports genetically engineered crops like an African sweet potato with five times the nutritional value of a natural potato, golden rice with beta-carotene or vitamin A, and soy richer in protein (Advantage, 2001). Although golden rice does not yet exist in commercial form—it is a research crop produced without profit by the Rockefeller Foundation—it is estimated that the strain will annually save one-half million children from blindness and two million children from anaemia.

An equally impressive and odd display of biotech advocates was seen on a segment of Jim Lehrer's PBS NewsHour entitled "Genetically Modifying Food: Playing God or Doing God's Work"? It included a panel discussion among leading religious leaders on the ethical considerations of modern agriculture. A professor at Baltimore Hebrew University, Rabbi Reisner, said that Jewish thought is "supportive of the active, innovative manipulation of the matter of our world for human betterment. The question when we contemplate the manipulation of the nature is…as God's emissaries, are we doing good, which is God's will, or…have we substituted our will such that we pursue that which is not good"(Frazier, 2001, 2). Similarly, the Vatican's leading expert on bioethics said that although Pope John Paul II has not made a definite Catholic statement on the morality of genetically altered organisms in food, he has told farmers that it is a disgrace that in the twenty-

[11] Please see widely circulated press releases by the U.N. Food and Agricultural Organization (FAO) and U.N. World Health Organization (WHO) dated March 05, 1998.

first century people were still starving; newly modified food crops had the potential to improve the lot of the world's poor; and biotechnology methods "must be placed beforehand under rigorous scientific and ethical controls to make sure they do not result in disasters for human health and the future of the earth"(Wooden, 2001, 2). Similar cautious but theological justifications of biotechnology as a cure for world poverty can be found in major Eastern world religions (Warren, 1999). Interestingly, despite rejecting technological innovations like automobiles and mechanized tractors that disrupt community unity and the accepted rural work ethic, Amish farmers in America endorse genetically modified crops to achieve global humanitarianism and to secure their religiously strict way of life (Strawley, 2001).

As the human population continues to increase and the amount of arable land continues to decrease, many experts contend that food biotechnology is the best solution because it can feed more, better, and with less finite resources, thus minimizing the impact of agriculture on the environment. Advocates contend that not only is biotechnology needed in the future, it is, in fact, the low-risk alternative to current policies. Proponents maintain that, "[b]y reducing dependency on chemicals and tillage through the development of natural fertilizers and pest-resistant plants, biotechnology has the potential to conserve natural resources, prevent soil erosion," improve waste treatment practices, and "offer new ways to use renewable resources for materials and fuel"(McGlouglin, 1999). For instance, the USDA has convincing statistical data that explains that two million fewer pounds of insecticide were used in 1998 than in 1995 because of the use of genetically modified corn. In fact, all three major federal agencies—the USDA, EPA, and FDA—have released documentation that food biotechnology will lead to fewer riskier practices because it replaces expensive and toxic chemical inputs with expensive, but apparently benign genetic information while enhancing food quantity and quality.[12]

In addition to bureaucratic endorsements, organized religious support, international government advocacy, and major university studies, non-governmental organizations also support biotechnology as a tool that could help solve not only world hunger but also environmental contamination. For instance, in 1991, the Environmental Defense Fund—now simply Environmental Defense—believed that food biotechnology had the potential to solve more environmental problems than it could ever cause. Therefore, in order to gain support of less conservative environmental groups, it proposed

[12] Please see Federal Register (1992), FDA (1995) and read the Congressional Research Service Summary, 1999. "Food Biotechnology in the United States: Science, Regulation, and Issues." June 02.

that the FDA reconsider its position that genetically engineered ingredients are not inherently food additives (Martineau, 2001). The FDA's strictest regulatory controls and most stringent safety testing in any category is for food additives, and because Environmental Defense believes that most biotech foodstuffs would pass the internal tests of the FDA, it should require them to demonstrate the safety of biotech foods with exemplary data that would concur with the industry tests that many environmental groups consider suspect.

According to other nongovernmental organizations—like medical associations, for example—the public must be informed of the scientific benefits of biotechnology and its novel food products. For instance, in addition to the ruling body of the FDA, the World Health Organization, the American Medical Association and the American Dietetic Association have concluded "there are no food safety or health concerns related to BST residues in products such as meat and milk from treated animals" (Monsanto, 2001b). To combat irrational fears and unscientific risks of biotech foods that saturate the media, some groups have published the safety data on biotech milk, cheese, corn, and soy in scientific journals or made the results of such tests available to the public through the Freedom of Information Act (Martineau, 2001). The American Dietetic Association, for instance, released external scientific studies debunking the theory that biotech crops engineered to be immune to chemical pesticides could create superweeds—undesirable plants made resistant to herbicides because of genetic drift. Many other organizations with no apparent stake in biotechnology trumpet the notion that humankind can greatly benefit from genetically modified food products. A good example is the non-profit Jimmy Carter Center. Writing for his charitable organization, President Carter states:

> In the past 40 years, farmers worldwide have genetically modified crops to be more nutritious as well as resistant to insects, disease and herbicides. Scientific techniques developed in the 1980's and commonly referred to as genetic engineering allow us to give plants additional useful genes. . . Genetically engineered cotton, corn and soybean seeds became available in the United States in 1996, including those planted on my family farm...The number of acres devoted to genetically engineered crops in Argentina, Canada, Mexico, and Australia increased tenfold from 1996 to 1997. The risks of modern genetic engineering have been studied by technical experts at the National Academy of Sciences and World Bank. They concluded that we can predict the environmental effects by reviewing past experiences with those plant and animals produced through selective breeding (Carter, 1998).

While some groups endorse a stronger regulatory framework than others for maximizing the safe use of new agricultural technologies, virtually all of them seem to argue that building public assurance in and support for biotechnology is an essential tool in the sustainable development of modern food production.

5.LOOKING FORWARD

In the next two chapters, we will explore how the so-called objective evaluations of risk assessments conducted by dominant industries and governmental administrative bodies are, in fact, value-laden. Specifically, it will be argued that the technical rules, strategies, and methodologies used in risk assessment are similar to the more subjective process of risk management. According to the National Research Council, risk assessment is the "characterization of the potential adverse health effects of human exposures to environmental hazards" and risk management is "the process of evaluating alternative regulatory actions and selecting among" risk assessment studies for the purposes of ensuring human safety and environmental integrity (Council, 1983, 18). Whereas "risk assessment presumably is a scientific procedure which 'uses the factual base' to define the health effects of exposure of individuals or populations to hazardous materials and situations," risk management is presented as a completely separate bureaucratic function that is intentionally biased in order to produce certain public policies and regulatory rules (Cranor, 1990 ,125).

Risk management has three objectives: "1) reducing risks by reducing the probability of the occurrence of adverse events, 2) spreading risks across a group such that particular individuals or subclasses are not inequitably subject to noncompensated risk, and 3) allocating responsibility for compensation to those producing risks"(Sergerson & Bromley, 1985, 101; Thiele, 2000, 544). However, "the supposedly objective scientific studies used for estimating risks to human health for regulatory proposes can be considerably more controversial and political than most people think"(Cranor, 1990, 126). Most simply, there is a significant difference between the exacting scientific standards of evidence for risk assessment and the less demanding standards used for predicting regulatory estimates of harm in the same evaluation. For example, although the science used to determine the levels of carcinogens found in a food additive is very strong, the scientific probabilities about the amount of the additive to be consumed by healthy human populations in the government's final risk assessment are a lot weaker.

Risk analysis, despite its necessity and use, has been criticized by technical elites and informed citizens for its lack of scientific rigor and its failure to predict and evaluate environmental and health impacts of a proposed course of action with any real certainty. For example, some critics of the current rational paradigm have argued that because of the small number of samples of rare diseases used in epidemiological assessments, and the scientific method that urges studies be 95 percent certain that statistical inferences from sample populations do not produce false positive results, researchers might be failing to discover many significant health risks (Cranor, 1990). It has been stated that the "management of human activity to minimize adverse environmental effects depends on information and understanding about ecosystems function which is invariably limited"(Treweek, 1996, 196). Regardless of the suspect methodology and limited scientific data, regulators are under pressure to develop models to determine the uncertain long-term effects of new products and processes without any certain multivariate correlation data of short-term effects between environmental impact and public health.

The choice of variables used and the statistical methods employed in the regulatory science of assessment are based on moral and policy judgments indicative of the subjective politics of risk management and not the so-called objective science of risk assessment. This is important because often scientists in risk assessment interpret their studies for risk managers and consequently determine "substantial environmental public policy under the guise of science without public scrutiny"(Cranor, 1990, 138). "In a democracy, it is arguable that such important policy issues should not be hidden in the science but settled in the more democratic forum of regulatory hearing by risk managers who are to some extent accountable to the public"(Cranor, 1990,138). Risk assessment is a very inexact science. Yet, many administrators and public policy planners who make decisions about health polices and environmental programs operate with a certainty that is more a product of their moral and political philosophy than the guidelines offered by risk estimates based on their scientific data.

As massive environmental problems are burgeoning before us, public policy discourse will become more important. Although these problems are scientifically complex, their political solutions should not be dominated by technical language. Recognizing that participants in plural political processes bargain, given their interests, but also refine and learn from others' interests, risk assessment and management should look at participatory policy models and post-positivist perspectives on policy studies (Fisher & Forester, 1993). That is, even though mathematical programming, cost-benefit analysis, and probabilistic risk assessment will remain important techniques in weighing environmental risks, other approaches need to be considered in order to judge all relevant perspectives. The policy sciences are oriented towards a positivist design, with economics and many of its applied methodologies serving as the dominant approaches. These models, however, have not been serviceable in explaining the importance of democratic inclusiveness in policymaking. The next chapter explains how risk assessment is based on the utilitarian principle of the greatest good for the greatest number, but in the end relies on a rational choice methodology that seeks to find the lowest risk for the greatest number.

The three emerging theoretical frameworks for contemporary policy studies are structural choice theory, institutional rational choice doctrine, and advocacy coalition framework. All three of these approaches only consider policy discourse in a closed rational fashion and not in an open, communicative, and liberal manner. For example, all of these methodological perspectives maintain that the sole objective of risk management is safety control for the consumer and not the building of trust in the bureaucratic process for the citizen. This will be best understood in Chapter Four, which describes risk management as exclusive administrative decisionmaking rather than an inclusive social contract that involves public consent or trust in the power of administrators who make representative decisions for the polity.

Finally, in the next couple of chapters it will be argued that these dominant administrative approaches of assessment and management are subject to an ideological bias that favours new and contested technologies (Plutzer, Maney, & O'Connor, 1998). In fact, it is not surprising that policy frameworks that emphasize scientific knowledge, expert technical opinions, and elite policymaking will often favour economic innovation over sustainable and cautious development of emerging technologies. As a result of this bias, however, many environmental policy scholars advocate democratic influence as a necessary but not sufficient condition of good public policymaking (Thiele, 2000, 559). It is essential to note that in the assessment and management of food biotechnology, this democratic discourse should accompany, not replace, scientific and technical communication of environmental risk policy (K. Shrader-Frechette, 1985).

In the technologically advanced and economically aggressive American society, environmental and human health risks are unavoidable. Some liberal theorists, like Immanuel Kant, argue that human agency, dignity, and sanctity transcend all calculations of price and any policy discussion that would determine the value of human life is unconscionable and immoral (Gillroy, 1993; Thiele, 2000). However, that ethical position is not reasonable in a complex society of personal and public risk tradeoffs. That is, while it might be easy to stop drinking saccharin-sweetened diet soft drinks with the knowledge that you decrease your chance of cancer by one in a million for every 30 cans you do not consume; it is difficult to eliminate breathing New York City's polluted air with the knowledge that in doing so for two days decreases your chance of death by one in a million (Thiele, 2000; R. Wilson, 1990). It does not seem unreasonable, however, when liberal theorists like Locke and Hobbes assert that risks should be reduced and managed by an informed public. Furthermore, while risk assessment and management are prone to benefit the producers of risks rather than their victims, they should not be dismissed as tools of the master, but critically engaged as an apparatus of transformation in order to value otherwise ignored concerns (K. Shrader-Frechette, 1985, 81 and 200).

It is important to note that injecting liberal political philosophy into public policy deliberations is not a new concept. For instance, more than a century ago, Norton E. Long challenged the idea—entrenched in orthodox public administration theory—that public bureaucracy could divorce its administrative functions from political theory (Long, 1949). Long argued convincingly that public administrators must engage in politics and power seeking to accomplish their objectives. Today, public administrators have adapted and have developed their own power base through alliances with interest groups and key members of Congress. Public administrators continue to develop this power base by looking to untraditional political actors. In the case of food biotechnology, governmental risk managers must consider industry studies, university advocates, environmental protestors, policy stakeholders, and political facilitators.

Similarly, James Q. Wilson argues that American bureaucracy faces several problems—accountability, equity, efficiency and political responsiveness—which require solutions that may be mutually incompatible (J. Q. Wilson, 1967). By suggesting that in order to solve the problems of bureaucracy we must first re-examine fundamental public goals and clarify programmatic objectives, Wilson is arguing that public administration must be directed by political theory. Like Wilson, I am simply arguing that the dominant positivistic perspectives of risk policy analysis have incorrectly dismissed questions of values and ethics (W. Dunn, 1983). As such, I seek to correct this deficiency by addressing social values and analytical ethics among risk analysts and public managers in a way that is different from traditional economics and science. This book addresses the defects in the current rational-scientific reasoning used in risk analysis and proposes an alternative by suggesting a political theory approach to understanding the social-liberal questions of environmental value in food biotechnology.

The post-positivistic policy team of Frank Fisher and John Forester is also concerned about the discourse among policy planners and the public. They seek to synthesize the two contrasting tendencies in public policy analysis—the rationalistic problem-solving model and the politicised context-determinant model—in order to create a combined field that is to be neither methodological nor contextual alone (Fisher & Forester, 1993). In order to avoid the separation of epistemological concerns from institutional concerns, they focus on argumentation and compromise as the locus of policymaking. Similarly, I hope to broaden the discourse base of risk policymaking by looking at food biotechnology as a political deliberation and not merely a technical one. The next two chapters will discuss expanding the liberal values of administrative rationalism because institutionalising a political theory focus in the practice of risk assessment may achieve the necessary liberal equilibrium between assessment and management.

CHAPTER THREE

THE UTILITARIAN ASSESSMENT

The Politics of Living with the Smallest Risk for the Greatest Number

1.INTRODUCTION

To the inquiry: "...do you think that the first thing we should aim at is to take measures to avoid danger?" Socrates responds, "Certainly not"(Plato, 1974, 128). To the rebuttal question, "Well then, if risks must be run, should one not run them where success will improve people"(Plato, 1974)? Socrates answers without hesitancy, "Obviously" (Plato, 1974). Thus, the modern notion that societal risks should be taken for the utilitarian good of the political community is embedded in one of the classical colloquies of Plato's *Republic*. In fact, historians have traced the use of professional risk assessors to an earlier time in ancient Babylon. In the year 3200 B.C., it is recorded that people calculated the communal utility in risky decisions like marriage proposals and construction sites (Boroush, 1998). More importantly, for more than a quarter of a century in American public policy, some version of the philosophical notion of utilitarianism—"the modern theory that judges the goodness of outcomes and therefore the rightness of actions. . .by the degree to which they secure the greatest benefit to all concerned"—has been the foundation for assessing environmental risk (Hardin, 1988, xv).

This chapter examines the gap between the modern philosophy of utilitarianism and the contemporary methodology of rational choice in risk assessment. The first systematic proponent of utilitarianism, Jeremy Bentham, is used as the basis for a comprehensive theory of public policy. Bentham's utilitarian practice of making law and policy based on the desired and unwanted consequences of alternative proposals is easily compared to the rational choice theory of identifying the optimal choice in political science. Initially, I discuss the extent to which utilitarianism offers a complete structure for liberal principles, including respect for human rights and environmental integrity. Second, I focus special attention on the challenges posed by using only rational choice logic in democratic decisionmaking. Third, I reason that environmental risk assessments based primarily on Bentham's utilitarianism are a process of subjective rationality that politically distorts the more inclusive utilitarian liberalism of Bentham's best student, John Stuart Mill. Finally, using some critiques of Mill, I argue that risk issues associated with agricultural biotechnology are a paradigmatic demonstration of the defects of using only a rational choice methodology for defining ecological harm. That is, while the "probabilistic assessment of risk" is helpful in determining a decision rule, it must not be the only indicator used in accepting or rejecting a risk in environmental public policy.

"Risk. . .involves the chance of loss or the probability of ruin as well as the alternative chance of some payoff that exceeds the investment in scarce resources"(Deibert, 1990, 131). Environmental decisionmaking, more than any other policy area, is plagued by unequivocal evidence, social ignorance, economic uncertainty, and moral implications. Waiting to overcome public ignorance or discern scientific cause and effect is not viable as political delay may cause further damage to the concerned ecosystem, genetic diversity, or polluted area (Handmer, Norton, & Dovers, 2001). Environmental decisions must be made with uncertainty but are formulated with predicting the dangers, measuring the risks, and pondering the consequences of a policy action. Therefore, understanding the magnitude of danger posed to human life, public health, and affected ecosystems are at the foundation of most environmental programs.

In order to decide whether to undertake an action when operating with the knowledge that success is uncertain and that unintended consequences may happen, the process of risk assessment is used. It is similar to cost-benefit analysis—the simple calculation of "pluses" and "minuses"—except that in risk analysis the "minuses" incorporate measures of the likelihood of negative effects as well as of their magnitude (Stone, 1997). The qualitative results of this rational equation are hardly ever reported. That is, you seldom read governmental reports that use terms like low, medium, or high risk, because "risk is frequently expressed in quantitative probability terms—such as some number of additional cancer deaths over a lifetime in a population of 1 million exposed people" (Boroush, 1998, 8). For example, "a risk of 1 in 10,000 people is often described as a '10^{-4} risk', 1 in 1 million as a '10^{-6} risk', and so on" (Boroush, 1998, 8). Societal costs and benefits of a policy action predicted to have a '10^{-4} risk' would be weighed by policy experts, but risks of less than 10^{-6} magnitude are not the object of concern for industry risk assessors and consequently governmental regulators.

Risk assessment is a controversial set of methods used for estimating both human health and ecosystem risks (Byrd & Cothern, 2000). It has been said that the calculation of cost estimates in risk-benefit analysis is simultaneously straightforward and overwhelming (Stone, 1997). That is, the procedure is straightforward because it is based on multiplying a measure of the magnitude of an outcome, such as dollars or lives lost, by a decimal fraction that expresses its probability of occurring (Stone, 1997). Stone agues, it is at the same time overwhelming because the technique builds in an assumption that a bad result is less bad if it is less certain to occur. This predictive nature of risk—the probability factor that will create the expected value—is important because it often determines the final environmental policy decision. Environmental regulations in the United States that assess risk as the probability of future loss include standards for particulate matter in air, arsenic in water, and pesticides in food (Byrd & Cothern, 2000). The specific methods for conducting a risk assessment vary according to the context of the policy action and whether chemists, biologists, toxicologists, epidemiologists, engineers, or environmental scientists are conducting the study. The general processes of assessment, however, consist of a set of four independent steps.

The incremental steps of risk assessment—source assessment, exposure assessment, dose-response evaluation, and risk characterization—are worth explaining as a basis for understanding the process. At the outset, source assessment identifies the sequence of hazardous events through which an exposure to risk could arise. This first step of risk assessment is conventionally used to identify logically possible exposures to danger from new processes or products. Engineers who assess the safety hazards associated with the typical uses of an appliance employ this kind of analysis when they decide what warning labels to include with the new product (Boroush, 1998). Secondly, exposure assessment determines "the likelihood, intensity, and duration of human exposure to the hazard identified, taking into account the worst-case scenario or the reasonable maximum exposure (RME)"(Armour, 1997, 39). An example would be using data from current, past, or present epidemiological studies of chemicals that could be found in a dump in order to estimate the total possible exposure of a hazardous waste landfill. Perhaps the most common use of this assessment method is to determine how much a new substance can be used before it threatens human health or a functioning ecosystem. This type of assessment is then used to determine how close a hazardous waste landfill can be to a public water source without a probable contamination.

Thirdly, dose-response evaluation is a detailed study conducted to discern what relationship exists between the level of exposure and the incidence of adverse effects. The toxicological analysis of the chemical Alar and its carcinogenic breakdown substances found in apple products consumed by laboratory rats is an example of this evaluation method (Henderson, 1993). As a result of most dose-response research being conducted on non-human subjects, evaluations must make an effort to simulate the effects on the human population without any direct data for this assessment. Finally, risk characterization is the concluding step for most technical experts doing risk assessments. "The total exposure estimate calculated in the exposure assessment is multiplied by the potency estimate developed in the dose-response evaluation to derive a final estimate of the risk for a given population" (Armour, 1997, 39-40). In the hazardous waste landfill case, the mathematical risk of each discarded chemical would be combined to obtain a probable estimate of the total risk for the entire landfill (Armour, 1997, 39-40). The results of risk characterization are used to calculate the final cost-benefit report to be used in discussions of program implementation by policymakers and stakeholders

Industry and administrative technicians involved in risk characterization maintain that sound environmental policymaking is (1) mostly an analytic, rather than political, enterprise; (2) works best when risk is measured in terms of expected losses—for example, projected deaths and/or injuries—because this is how the policy analyst conceptualises problems; and, (3) is achieved when risk values are reduced to a common metric to compare, trade-off, and aggregate environmental policy (Hornstein, 1995, 149). Expert risk assessments that use the "hard version," "technical understanding," or "standard account" of rational choice theory are said to be scientific, objective, and value-free. Producing risk evaluations on such hypothetical-deductive lines is endorsed by many of the mainstream figures of risk studies: Starr, Whipple, Maxley, Cohen, Lee and Okrent (Margolis, 1996; K. S. Shrader-Frechette, 1991). In fact, it has been correctly stated that evaluating risk by rationally calculating its probable consequences against its benefits is so customary to public policy that it is difficult to imagine assessments done without this utility calculus. The empirical method of reductive analysis used in risk assessment—calculating the objective criteria of benefits, pleasure, happiness, or goodness besides the contradictory criteria of costs, pain, unhappiness, or evil—is generally attributed to the philosophy of utilitarianism.

Utilitarianism is an ethical theory with political consequences, and, in its standard nineteenth-century statement directs individuals to produce the greatest happiness for a society (R. Goodin, 1995, 3-4). Its most celebrated author, John Stuart Mill, best explains the political theory of utilitarianism in his "Essay on Bentham," Mill writes:

> It will enable society...to prescribe the rules by which it may protect its material interests. It will do nothing...for the spiritual interests of society...what a philosophy like Bentham's can do [is to] teach the means of organizing and regulating the merely business part of social arrangements (Mill, 1962, 104-6).

Since its inception two hundred years ago, the approach has matured philosophically, resulting in, many diverse forms of utilitarianism.

"Philosophical debate has questioned whether the [utilitarian decision rule] should be a maximizing one, a libertarian one, a game-theoretical one, a distributive one or some combination of rules; but all these debates take place within the policy framework advocated by Bentham"(P. Thompson, 1990, 2). While these internal distinctions and debates are important, the fact that utilitarianism is the foundation for American public policy is what is of real significance for this discussion.

Indeed, the original utilitarian calculus has become the predominant model for developing public policy and the very interpretation of rationality for political science (Friedman, 1996). Commonly referred to as the "rational choice theory" it describes how individuals should make consistent, value-maximizing, optimal choices. Best choices are achieved by following a rational decisionmaking process that "(1) defines the problem, (2) identifies decision criteria, (3) weighs the criteria, (4) generates alternatives, (5) rates each alternative on each criterion, and (6) computes the optimal decision"(Robbins, 1997, 77). In addition to creating a model for developing optimum policy preferences for a social body, advocates of utilitarianism argue that such a rational decisionmaking process is an ideal fit for democratic regimes using a system of electoral representation. For instance, Mill once said that Bentham has "pointed out with admirable skill the best means of promoting one of the ideal qualities of a perfect government—identity of interest between the trustees and community for whom they hold their power and trust"(Mill, 1950, 88). Such an argument is so commonplace that undergraduate American government textbooks explain how the utility principle or individual moral decisionmaking equation is used as a system of democratic public policy. An example is as follows:

> Governments should choose policies resulting in gains to society that exceed costs by the greatest amount, and governments should refrain from policies if costs are not exceeded by gains. Note that there are really two important guidelines included in this definition of maximum social gain. First of all, no policy should be adopted if its costs exceed its benefits. Second, among policy alternatives, decision makers should choose the policy that produces the greatest benefit over cost (Dye, 1992,30-31).

The basic utilitarian approach to creating social policy has not changed since Bentham and is easily seen in public policy discussions that focus on achieving maximum social gain or question whether societal risks are politically rational. Bentham, although betrayed by contemporary utilitarianism, is credited with employing the scientific method or quantitative analysis to make decisions in politics, law, and justice.

2.HISTORICAL DEBATES WITHIN RISK ASSESSMENTS

Essential to any discussion of risk is an appreciation of the political controversy surrounding scientific methods of assessment. On one side, environmentalists have been sceptical of relying on technical evaluations out of fear that the severity of ecological dangers might be understated when subjected to a scientific analysis. Jack Doyle's book *Altered Harvest*, for example, argues that the "generally recognized as safe (GRAS)" category of hundreds of food additives that emerged out of the 1958 amendment to the Food, Drug and Cosmetic Act is a perfect illustration of a technical evaluation of safety that reduces many environmental concerns about food risks (Doyle, 1985, 147). Industry, government and scientific experts—on the other side—often claim that many risk assessment methods and related policy decisions are too conservative. Chauncey Starr, for instance, an engineer and advocate of nuclear power, has for years criticized dose-response evaluations that cautiously calculate levels of radiation under the assumption that individuals will eat, drink, and sleep at the very fence of a power plant. Of course, both opposing characterizations of risk assessment are accurate to some degree, because risk assessments are a useful, albeit imperfect, means to reach conclusions required for environmental decisionmaking. That is, risk assessments have the potential to establish a formal process that can bring into open debate many of the otherwise hidden economic, social, and technological biases that influence American environmental public policy. Such a process creates a space for debating different political values.

As explained, the technical community defines risk as a product of calculating the probability of an event and the seriousness of the consequences, such as health or environmental effects, which can follow. Risk connotes an indefinite problem, occurrence, or exposure that can be described only in terms of mathematical probabilities rather than certainties. The public community, on the other hand, views risk differently and gives greater consideration to qualities, such as the degree to which risks are uncertain, uncontrollable, fatal, or catastrophic. As a result, the public's perception of risk often differs drastically from that of scientists, engineers, and governmental regulators. These differences, however, complicate the utilitarian conduct of risk assessments because, although technical evaluations are just as subjective as public perceptions, they often are presented as objective scientific determinations. Despite this confusion, credible risk assessments that are scientifically sound and socially accepted are needed to manage American society's environmental problems and technological advances.

In the early 1970s, the science of risk analysis was created in order to quantify the effects of proposed technological projects and products. This development was a direct consequence of the U.S. National Environmental Policy Act (NEPA) that required all federal agencies—such as, the Occupational Safety and Health Administration (OSHA), the Department of Transportation (DOT), the Nuclear Regulatory Commission (NRC), the Department of Energy (DOE), and the FDA—to prepare environmental impact statements (EIS) for all federal activities that could affect the environment (K. S. Shrader-Frechette, 1991, 6). Many risk scholars, however, complained about the methodological inconsistencies of these reports. For example, the automobile industry determined that the life of every passenger saved was worth $30,000 while the steel industry decided that an expenditure of $5 million was equivalent to the life of each steelworker saved (K. S. Shrader-Frechette, 1991, 6). Such inadequate standards and regulatory difficulties prompted Congress to pass the Risk Analysis Research and Demonstration Act (RARADA) to standardize the methodologies used for protecting human life, public health, and natural resources.

Many environmental scholars and activists contend, however, that RARADA never truly perfected risk assessment. The traditional cost-benefit framework prescribed by risk analysis is criticized because (1) its efforts to combat a target risk unintentionally foster countervailing risks; (2) its hazardous evaluations often contradict each other; (3) scientists disagree regarding relevant facts for making an analysis; and (4) the affected public often disputes technical judgments of risk (Graham, 1995). These criticisms are partially correct because risk assessment takes place within a context of societal uncertainty and not in a scientifically controlled laboratory. This contextual problem, however, does not mean that the underpinning theory of utilitarianism in risk assessment needs to be rejected. Rather risk analysis only needs to be augmented by an inclusive consideration of some of the neglected environmental, societal, economical, and political ethics involved. Although involving environmental ethics, this project is a more comprehensive investigation of the political theories that might inform the current practice of risk analysis. There is a disturbing tendency in risk scholarship to imply that environmental ethics is a branch of applied ethics—used to establish moral principles for environmental decisionmaking and the formulation of sustainable and just environmental policy (Cooper, 1998, 196). This implication needs to be refuted because "rather than applying moral theory, environmental ethics has been primarily devoted to changing" some of the problems with the dominant economic, societal, scientific, and political paradigms (Cooper, 1998, 196).

The foundational thesis of this work is that there are strong correlations between political theory and public policy; primarily that the former is a foundational guide for the latter. However, it is not my conclusion that Jeremy Bentham's utilitarianism is the theoretical structure for the contemporary practice of risk assessment. For, it is not useful to declare that the theoretical underpinning for contemporary American environmental public policy is utilitarianism. In the first few pages of *Introduction to the Principles of Morals and Legislation,* for instance, Bentham himself states that he is creating a utility theory of law to comprehensively guide public policy. Instead, what I am arguing is that there are philosophical limitations and unreasonable consequences of having Bentham's ideas undergird the values, strategies, and objectives of environmental risk assessment. Moreover, those consequences could be better observed and possibly altered if risk assessors were to look at the liberalism of John Stuart Mill and the scientific philosophy of Sandra Harding.

The following descriptive analysis, therefore, has prescriptive consequences. Whereas a utilitarian discussion aids in the theoretical understanding of risk assessment, it also serves as a practical guide or normative proposal for experts concerned with the problems emerging from environmental practices. The significance of this work is found in discussing the limitations of Bentham's political system and explaining the responsibility and ethical burden its impoverished utilitarianism bears. Thus, in augmenting Bentham's utilitarian calculus with Mill's notion of social excellence and Harding's critique of objective epistemology, risk assessors may better understand why rational definitions of risk have social critics. And, by identifying the theoretical limits of dominant risk-benefit assessments, risk managers might be able to avoid, mitigate, or compensate for some of their problems in practice.

3.BENTHAM AND THE UTILITARIAN ETHIC IN RISK STUDIES

As noted, rational risk assessment's practice of only using quantitative methods is indebted to Jeremy Bentham's political philosophy of utilitarianism. According to Bentham, the tasks of government should include the following: (1) maintain an organized, consistent, and rational legal system; (2) secure the individual wealth and property of citizens; (3) cultivate a spirit of kindness among its public; and, (4) "although economic, political, intellectual, and moral inequalities are both unavoidable and necessary, government should moderate their extreme forms" by the "principle of utility and based upon the respect for secured property"(Wiser, 1983, 294). Bentham, who writes more like a political reformer than a systematic philosopher, has two works that explain his theory of utilitarian ethics: *Fragment on Government* (1776) and *An Introduction to the Principles of Morals and Legislation* (1789).

According to Bentham, the subjective judgments of moral or social sciences should be rectified by objective measurements of the physical or natural sciences. For instance, the "fictions" of political theory—justice, power, and rights—must be reduced to their empirical objects or base emotions. Bentham saw himself as developing an empirical science for personal virtue and public morality. Arriving at an impartial, value-free, or objective foundation for ethical reasoning was an important political aim for Bentham. This is nicely illustrated by Bentham when he writes the following:

> The Community is a fictitious body, composed of the individual persons who are considered as constituting as it were its members. The interest of the community then is what?—The sum of the interests of the several members who compose it (Bentham, 1967, 123).

According to Bentham, everything can be explained in utilitarian terms. For instance, a communal or political good is simply the mathematical aggregate of all of the individual goods of its members. Similarly, for Bentham an individual good or human happiness is defined solely by the accumulation of pleasure and avoidance of pain.

As a moral philosophy, utilitarianism straightforwardly claims that human happiness is a mathematical computation in which the pluses of pleasures exceed the minuses of pains. Since human happiness is the highest good, all ethical questions are determined by calculating which option yields the greatest amount of human happiness or pleasure. Bentham famously writes:

> By the principle of utility is meant that principle which approves or disapproves of every action whatsoever, according to the tendency which it appears to have to augment or diminish the happiness of the party whose interest is in question (Bentham, 1967, 126).

As a political system, utilitarianism argues that government must determine the amount of pleasure its various policy options will produce and decide on what means are available to implement these policies. "In deciding which policies it is to pursue, the government must calculate" each program's ability to "augment or diminish the pleasure of the total amount of individuals concerned"(Wiser, 1983, 300).

Bentham maintains that determining public pleasure is purely an arithmetical operation that involves seven external circumstances and thirty-two internal factors that influence individual pleasure. Because of its prolonged and difficult calculation, Bentham suggests that this process "be always kept in view" but "not strictly pursued to every legislative or judicial operation"(Bentham, 1967, 153). Implementing its policies and enforcing obedience among the citizenry, however, is simply achieved by the government granting pleasures or threatening pain with its preferred policies. According to Bentham, all human behaviour can be controlled by governmental sanctions of pain and pleasure. For Bentham, utilitarianism is the necessary and sufficient means for political order because it is the most reasonable understanding of power. "The Benthamite solution to the problems of political order exhibited a certain rationalistic and ultimately mechanistic quality"(Wiser, 1983, 305). The relationship between citizens and the government is expressed in terms of the individual's rational self-interest, and the justification of governmental power is its ability to facilitate the greatest amount of self-interest for the greatest number of people.

Bentham's hierarchically controlled formulation of a mathematical calculus which assumes pleasure and goodness to be synonymous is the model for current democratic decisionmaking. In fact, many philosophers refer to legislative policymaking as "preference satisfaction" which is a sophisticated form of the utility principle that allows for mistakes in rulemaking and considers human happiness as more than appetite and aversion. As mentioned, there are entire books dedicated to the subject of utilitarianism as the quantitative model for American public policymaking (R. Goodin, 1995). Risk scholars have already made the connection between Bentham and risk assessment. For instance, Thompson has written:

> For the Benthamite, all acts have consequences and most will have some bad that must be weighted against the good. Risks are not fundamentally different from negative outcomes, aside from the fact that their probabilistic character complicates our decision rule for weighing good and bad. Indeed, it is the consequentialist followers of Bentham in contemporary utility theory and rational choice that have given us the elaborate definitions of risk as a function of probability and utility of outcome. This means that there is no bias against risk per se (aside from the obvious fact that risks are counted on the "cost" side of the policy evaluation). This acceptance or rejection of a risk can only be made after a predictive analysis that identifies (however roughly) the expected consequences, and assigns a probability to them. This assessment of an action's risk does not introduce a prejudice against the act, for it is only after the decision rule has been applied that a risk is determined to be acceptable or unacceptable (P. Thompson, 1990, 3).

It is indeed reasonable to conceive of an outcome by evaluating possible consequences. It is my position, however, that environmental risk assessors are compromised by the fact that they only use the limited Benthamite calculus of projecting the self-interested consequences of pleasure caused by an event and its alternatives, rather than a fuller qualitative utilitarian design of social welfare. Additionally, Bentham's consequentialism is obscured, because despite technical confidence, Bentham's model is not objective; it has subjective biases and, therefore, makes political assumptions about what consequences are significant to measure.

If utilitarianism is to be the normative perspective of environmental risk assessments, Bentham's political theory of consequentialism would be better served by embracing John Stuart Mill's liberalism that strives for social excellence. For example, the Benthamite goal of the penal system is to punish individuals who have acted in conflict with the societal happiness of a community. This is different from Mill who argues that while the primary goal of punishment is to deter, reformation of offenders should also be implemented because it can promote or harmonize the societal interests that create happiness. However, even with Mill's correction of social excellence, the prominence of objectivity in Mill's work hinders it from allowing rational actors to make mistakes or permitting happiness to be defined by subjective preferences. That is, as an empiricist who only evaluates the probable consequences of actions, Mill is unwilling to consider subjective influences that will create errors in calculating the aggregate preference satisfaction for a political community.

Unfortunately, most risk studies that discuss environmental ethics fail to notice the dangers of consequentialism or the powers of empirical-objectivity when using utilitarianism as a public philosophy. Perhaps the best examples of this danger are seen when scholars have applied utilitarian ethics to risk assessments of food biotechnology and particularly in the case study of bovine growth hormone. The most comprehensive examination is Ben Mepham's ethical matrix derived from the work of Beauchamp and Childress in *Principles of Biomedical Ethics*. Mepham maintains that his discussion of social ethics "will guide the implementation of necessary regulations on the employment of biotechnology in food production"(Mepham, 1996, 104). Seen in Table 3, Mepham argues that four interests—treated organisms or cows, producers or farmers, consumers or the public, and biota or the environment—are relevant in assessing the risk of using rBST.[13] Additionally, Mepham outlines the considerations of three traditional fields of ethics: well-being or the utility calculus; autonomy or Kant's deontological theory; and Justice or Rawls's fairness principle.

Assessing the ethical impacts of rBST to be used by decisionmakers involved in risk policy and as a model for scholars interested in the politics of contested technologies is an important project for which Mepham should be applauded. Mepham's table of the ethical impacts of rBST is not meant to be prescriptive; rather it is only a descriptive analysis of the moral issues that are relevant to policy evaluation.

[13] Ben Mepham's *Food Ethics* (Mepham, 1996, 106) and Ben Mepham's "A Framework for the Ethical Analysis of Novel Foods: Ethical Matrix" (Mepham, 2000, 170). Comments in parenthesis are largely taken from the many scholars who have used his matrix, especially Hayry (2000). For an interactive and more sophisticated version of the matrix please visit //http.www.ethicalmatrix.net.

Table 3. An Analysis of the Use of rBST with Mepham's Matrix

Respect for:	*Well-being*	*Autonomy*	*Justice*
Treated Organism	*Animal Welfare* (The use of rBST increases the incidence of diseases in cows and causes them discomfort.)	*Behavioural freedom* (Diseases and zero-grazing infringe the freedom of the cows treated with rBST.)	*Telos* (The intrinsic nature, or telos, of the cows treated with rBST is not respected.)
Producers	*Adequate income and working conditions* (The manufacturers and users of rBST can expect considerable profits. Other dairy farmers will lose.)	*Freedom to adopt or not adopt* (The autonomy of farmers who would not like to adopt rBST is undermined.)	*Fair treatment in trade and law* (The farmers who are not prepared to profit from bovine discomfort are unfairly penalized.)
Consumers	*Availability of safe food: acceptability* (rBST milk and decreased milk consumption can create health hazards and cause unease.)	*Respect for consumer choice* (Unless rBST milk is labeled, the freedom of choice of consumers is restricted.)	*Universal affordability of food* (The affordability of milk is not enhanced, as implicitly promised by the innovation.)
Biota	*Protection of the biota* (The use of rBST can reduce global pollution, but it can also intensify it both locally and globally.)	*Maintenance of Biodiversity*	*Sustainability of Biotic populations*

For Mepham, the purpose of such a matrix is to facilitate better participation for a broader debate about new biotechnologies because economic and scientific concerns traditionally outweigh ethical concerns. The central problem with Mepham's ethical perspectives on food biotechnology is that risk planners will only consider the consequences of rBST, and not question the technology's complete social utility or allow for subjective preference satisfaction. Although Mepham identifies the ethical concerns of different actors and frameworks, in the end his analysis is used to address only the consequentialist morality of a probable action.

Despite Mepham's statement that the motives of the advocates of a technological product are to be addressed in the autonomy and justice sections of his analysis, his final matrix suggests that all impacts can be calculated as if they could be quantified as commensurable costs and benefits. This interpretation can be seen when Mepham concludes that while "respect is shown for the well-being of manufacturers and some dairy farmers, and in some ways the environment... the effect on animals, many dairy farmers, consumers and the biota is negative"(Mepham, 1996, 111-112). Mepham seems to conclude that the deontological autonomy of consumers is lost because the respect for choice is contravened with rBST, but this cost is given the same weight as the loss of fairness in farmer trading. Mepham's matrix incorrectly implies concepts of right, wrong, and obligation rather than suggests how aggregate happiness can be obtained by allowing political actors to exercise subjective preferences.

Like Bentham, Mepham assumes all consequences can be measured by the aggregation of the interests of all self-interested parties. Although Mepham is not a consequentialist, his operationalized four principles of Beauchamp and Childress for biotech ethics are utilized as an ideal code of measuring the consequence of an action rather than deducing the context of an appropriate action. Granted Mepham does not contend that happiness is achieved by applying the hedonistic calculus. Unfortunately, his matrix fails to acknowledge that the attainment of individual happiness depends on the fulfilment of societal conditions for happiness. Mepham's matrix suggests that policymakers can inductively establish social happiness because they can calculate what will make all farmers and consumers happy. Although never easy to define happiness, it seems impossible to conceive that individuals could objectively reach that state when they are not allowed to subjectively consider their respective economic or political condition.

Regardless of Mepham's ambitious attempt to assess risks in light of multiple ethical approaches, decisionmakers will evaluates all of these perspectives as utility consequences and thus make his matrix a policy decision table. As with any innovative academic project, Mepham's matrix has been critiqued and revisited by many scholars. In order to address critics who have accused him of advocating a utilitarian ethic for risk assessors, Mepham states "the principle aim of the Matrix is to facilitate rational public policy decision-making by articulating the ethical dimensions of any issue in a manner that is transparent and broadly comprehensible"(Mepham, 2000, 169).

Despite characterizations of Mepham as a limited consequentialist, he steadfastly maintains that he is considering three principles of ethics, while admitting that it might be preferable to consider just the utilitarian principles of his analysis (Mepham, 2000). As such, one scholar, Matti Hayry, has used Mepham's ethical matrix to develop an environmental political theory of consequentialism (Hayry, 2000).

Working from Mepham's original matrix, Hayry attempts to use Bentham's utilitarian ethics to influence policymakers about the problems associated with rBST. Unlike Mepham, who offers an ethical framework for encouraging public participation, Hayry submits the following definitive normative statement:

> As long as dairy cows can be expected to suffer from the use of rBST, and consumers are subjected to health hazards by unlabeled rBST milk products, the use of the bovine growth hormone is harmful, and should therefore be banned. This means that the policy prevailing in the United States is unethical, and should be altered forthwith (Hayry, 2000, 181).

Hayry's systematic bioethical approach is evident when he admits that even though he is not an advocate of recombinant bovine growth hormones, if rBST could be administered to cows without causing painful side effects and if treated milk would be labelled, there would be no ethical grounds for rejecting its use.

As seen in Table 4, Hayry offers a clearer attempt at Bentham-like utilitarian analysis of rBST by distinguishing "well being" among interested parties as simply offering the benefits and harms of its use.[14] Whereas Mepham attempts to look at many ethical perspectives, Hayry admits his matrix will only be used by elites who will reduce the descriptive ethical impacts into utilitarian calculations. For Hayry, this is not a problem because he acknowledges that utilitarianism is the only ethical theory that will be considered in public policy analysis. Interestingly, Hayry anticipates the central criticism of his work by writing:

> A possible counterargument to my analysis is that by ignoring issues of justice and autonomy I have reduced Mepham's original model to a mere utilitarian calculus, thereby compromising its status as a matrix which should facilitate ethical decisionmaking of one's moral views. In a sense, this is perfectly true. I have tried to show how a person of a consequentialist persuasion would like to interpret Mepham's ideas, and this inevitably leads to the mitigation of factors like "respect for telos of cows." From my point of view, if cows suffer, that is bad. It they do not suffer from the violation of their "intrinsic nature" then I cannot see what the problem is (Hayry, 2000, 183).

[14] Please see Matti Hayry's "How to Apply Ethical Principles to the Biotechnological Production of Food" (Hayry, 2000, 182)

Table 4. How to Apply Hayry's Ethical Principles for Considering the Use of rBST

	Harms	*Benefits*	*Autonomy*	*Justice*
Treated Organism	The treated cows suffer from many ailments and from the extra-tight containment.			
Producers	The ethically oriented dairy farmers are economically harmed.	Ethically indifferent producers and users of rBST are benefited.	The ethically oriented dairy farmers are economically coerced.	Justice does not prevail.
Consumers	The users and refusers of treated milk products are subjected to health hazards.		If rBST milk is not labeled, the freedom of choice of consumers is curtailed.	
Biota	Intensified farming practices causes local pollution and may add to global pollution.	The decreased number of cows reduces methane gas releases and global warming.		

Hayry's matrix is a correct interpretation of Mepham's framework, in that Hayry admits that both frameworks reduce social welfare to a list of mutually exchangeable benefits and risks that will be evaluated in aggregate as if they were all equally significant consequences. Although dissimilar in approach, both Mepham and Hayry find the same ethical difficulties with rBST. This list includes some of the following: impacts on the well being of the treated cows, the disadvantaged position of producers who do not adopt the technology, and the lack of information and, therefore, control that the general public has because rBST milk is unlabeled.

While Hayry properly characterizes Mepham's work as consequentialism, he is unable to isolate the epistemological problem of objectivity in utilitarianism and thus advocates a similar matrix to Mepham's. Both frameworks are further complicated in that they fail to distinguish the ethical issues of risk management and communication from risk assessment. It seems that while Mepham wants to assess the social welfare of rBST, his work is as limited as Hayry's because he reduces everything to the assessment of risk and consequentialist morality. In order to counter Bentham's individual consequentialism and to understand the subjectivity of utilitarianism, we turn to Mill's notion of utility welfare as a normative framework for risk assessment.

4. MILL'S LIBERAL REFORMS

John Stuart Mill is considered the most influential utilitarian political theorist. Although he maintained an allegiance to most of Bentham's principles, he introduced a couple of philosophical reforms to the project. The two works most important for understanding Mill's political theory are his essay "On Liberty" (1859) and his book Considerations of Representative Government (1860). From these sources it is clear that many subtle distinctions can be made between the theories of Mill and Bentham, but the most important ones for the purposes of this discussion concern pleasure and control.

Bentham considers happiness exclusively in the quantitative terms as the amount of pleasures achieved and the amount of pains avoided. Mill maintains that there are qualitative differences among pleasures received and pains avoided. For instance, Mill states that intellectual pleasure is superior to sensual pleasure. This reform to the enterprise of utilitarianism is important because "by admitting that some pleasures are superior in kind to others, Mill is claiming that" when given a choice individuals may value the end of the pleasure, happiness, and not just the pleasurable means itself (Wiser, 1983, 307).

Second, while Mill is still committed to individuals maximizing pleasure and minimizing pain, and he remains steadfastly dedicated to the principle that an action is right if it brings the greatest amount of happiness to the greatest amount of people, he wants to create an individual with a greater potential for executing his or her life plan. Such a philosophical reform means that Mill includes the possibility that individuals pursue some things besides pleasure and the avoidance of pain as ends in themselves.

Furthermore, by insisting that individuals prefer intellectual happiness to physical happiness, Mill gives reason a privileged position in his good society or best polity. Thus, he sets up a hierarchy of pleasures or goods in the human condition that is particularly evident in "On Liberty," where he explicitly argues that society is capable of entering into a higher form of civilization characterized by individual freedom and intellectual growth. Mill writes:

> Human nature is not a machine to be built after a model, and set to do exactly the work prescribed for it, but a tree, which requires to grow and develop itself on all sides, according to the tendency of the inward forces which make it a living thing. . .To a certain extent it is admitted that our understanding should be our own: but there is not the same willingness to admit that our desires and impulses should be our own likewise; or that to possess impulses of our own, and of any strength is anything but a peril and a snare...It is not because men's desires are strong that they are ill; it is because their consciences are weak (Mill, 1988, 127).

For Mill, humans are an evolving species and calculations concerning happiness should consider human improvement, as Socrates insisted, and social excellence.

Mill's notion of governmental control of the many by the few corresponds to his vision of humankind as a progressive species interested in creating a higher form of civilization based on intellectual consensus. Recall that for Bentham control of individuals was achieved by universal sanctions of physical pleasure or pain. Actually, Bentham believed that obedience to legal norms was more likely to occur under the threat of pain than the promise of pleasure. Furthermore, Bentham maintained because citizens are only controlled by self-interest, popular self-rule was his preferred governmental structure. Quite simply, by deciding to punish certain acts, the government creates irrefutable democratic policy obligations because all citizens will individually refrain from undesirable social behaviours. Like Bentham, Mill believes that rights are derived from obligations, but his theory of governmental control is focused on achievement of individual pleasures rather than an avoidance of aggregated pain. For Mill, social progress is the result of intellectual development that allows individuals to question and challenge the most respected societal values, norms, and customs.

As a liberal, Mill respects the primacy of the individual and advocates a limited state power; however, it is within the state's power to carry out political programs that advance civilization. For Mill, good politics requires evolving policies for social excellence. Mill, for example, believed that political despotism was a legitimate form of government for those societies which had not yet evolved beyond the level of cultural barbarism. However, once its citizens had become capable of being improved by free and equal discussion, the despotic regime no longer served a progressive function and, according to Mill, should be replaced by a more liberal form of rule. Similarly, such liberal institutions and practices as constitutionalism and the recognition of political rights were appropriate techniques for securing intellectual progress in more modern societies of Europe. Yet, as a means to an end, their real value depended upon the efficiency with which they performed their function (Wiser, 1983, 311).

While concerned for the preservation of individuality, Mill accepts that a government is permitted to protect its citizens from danger and promote the greatest happiness of the greatest number. To that end, he endorses what is frequently referred to as the harm principle: the state is allowed to interrupt individual affairs for the sake of preventing harms to others. This secondary principle limits individual actions if their consequences will harm the primary goal of social utility and progress.

To Mill's mind, the mission statement of governmental programs and policies is promotion of individual progress on the basis of social order (Mill, 1988, 162). That is, liberty is essential for individuality and is necessary to develop social happiness and utility. He claims that individuality will lead to social good because where all pursue their self-interest, social utility will be maximized. John Stuart Mill is, then, an advocate of a governmental structure that values free choice and encourages a great deal of citizen participation. At the same time, although he is an advocate of self-government, Mill does not endorse the principle that all people should share equally in the exercise of political power. Mill feared that the American model of representative government may result in a weak democracy because it favoured the rule of the majority rather than minority interests, often intellectual achievements for creating the progressive future that Mill desired.

Mill can substitute Bentham's sense of democracy as popular self-rule with government by intellectual elite because he distinguishes between the act of governing and the act of controlling. For Mill, a democratic theory of government does not require that the power rest with the people, only that the people retain the power of controlling those who govern (Wiser, 1983). Mill, therefore, suggested that there be a popularly elected representative assembly that only serves as a forum for ideas, but also oversees and controls the professional elite by approving or rejecting their legislative actions. While Mill wants a democracy, he fears that political equality and majority rule will result in an egalitarian society that would fail to promote progress of the individual and, ergo, the community. Instead, Mill wants human operations of free, reasonable, and intellectual actions of self-direction. For Mill, the state has authoritative power because it is organized by principled individuals.

Principled individuals are necessary in Mill's formula because he believes that the state should always be evolving, and to do so it needs a force for creative change that will offset the countervailing forces of majority rule and political authority. Furthermore, he is concerned that a political authority should not infringe on individual rights, utilitarian means, or pleasurable ends. Mill seems to contend that although human beings are no more than the sum total of their parts—he still wants human freedom, toleration of others, and mutually respected rights—they should not be treated only as the sum total of their parts. Mill seems to imply that human rights are the starting point for discussing the nature of a political arrangement, regardless of the fact that no mention of these rights can be found in his philosophy. For Mill, all good arises from rationally autonomous agents who are capable of controlling their individual desires and deciding what is in the best interests of society. Although Mill's calculus for predicting social welfare is more sophisticated than Bentham's, it is important to question Mill's emphasis on elite agents who seem to function as objective scientists for the good of society.

5.THE RISKS OF MILL'S POLITICAL PHILOSOPHY

Mill's political philosophy of self-sufficient citizens enjoying self-government under a minimalist interpretation of the law seems to be similar to the aristocratic conservatism of Alexis de Tocqueville. After visiting the United States in 1831 to examine its penal system, Tocqueville composed *Democracy in America*. Aristocratic conservatives maintain that one of the eventual and undesirable consequences of commercialism, industrialism, and egalitarianism is a strong central government. According to Tocqueville, democracy might become as despotic and unstable as any monarchical or autocratic regime because the power of a citizen majority could produce social conformity rather than a society that respects individuality. Accompanying dangers would include creation of a political community defined by mediocrity rather than outstanding individuals, and an economic environment preoccupied with materialistic rather than welfare values. Because Tocqueville was suspicious that democracy may lead to Jacobinism—power being placed in the hands of incompetents—he offered some political suggestions.

Alexis de Tocqueville, like Mill, made an academic enhancement of a new practice of politics. John Stuart Mill and Alexis de Tocqueville are proponents of positive liberty, the freedom to accomplish certain human goals, because both offer suggestions to overcome the potential threats of negative liberty—freedom from social restraints that infringe on an individual's autonomy.

Conservatism, a response to the principles stemming from the French Revolution, rejects the idea of violent change and upholds the need for order and unity. Traditional conservatives believe that society, institutions, and methods of procedure have grown in an organic fashion and therefore, social institutions and traditions should be respected. Both Alexis de Tocqueville in *Democracy In America* and John Stuart Mill in "Consideration on Representative Government" separate themselves from the simple conservative camp of Edmund Burke because they not only criticize liberal politics, but also offer suggestions for change.

In *Democracy in America,* Tocqueville stated that in order to obtain liberal stability and to guard against the tyranny of the majority and the demands of authority, there must be institutions of local self-government, decentralized administration, and voluntary associations. Similarly, Mill endorses proportional representation, extending the franchise, and weighing votes in order to secure intelligence over majorities. This struggle between liberty and authority is at the very root of contemporary risk assessments based on a utilitarian ethic. More importantly, it is the Tocquevillian strain in Mill that has a specific few individuals making the calculations for the general welfare of the many.

According to Mill's utilitarianism, a policy begins with any concept of law that can be found in the practice of giving the greatest number of individuals the greatest opportunity for self and societal progress. Unlike common law that is based on practice alone, Mill thought public law should be based completely on the expected outcome of its legal practice (Ogden, 1993). More important, Mill argued that a governmental policy could be reasonable only if external political leaders evaluated its internal consequences. Thus, Mill proposed that societal leaders evaluate all potential public laws in order to determine the consequences for all affected parties to maximize aggregate utility (P. Thompson, 1998a, 120). Mill's "objective" design of having political decisions made independent of the society that they will impact is used by risk assessors today.

Policy analysts who calculate the decision to adopt rBST by weighing the expected economic value of the bovine growth hormone for the dairy industry against the human health and environmental risks of the new technology without actually communicating with the interested parties are a perfect example of Mill's aristocratic utilitarianism. The problem with this subjective notion of utilitarianism is that it reduces Mill's social welfare liberalism to Bentham's worst case of mathematical consequentialism. In fact, it is hard to imagine that Mill's comprehensive liberal enterprise could be used to endorse any political decision based on comparing such limited outcomes. It appears that Mill's philosophy, like Mepham's ethical matrix, has the potential for individual decisionmaking that allows rational actors to categorize preferences by considering different notions of satisfaction in order to create the most desirable social good. Unfortunately, industry and governmental risk assessors seem to forget that their utilitarian framework of evaluating choices is a subjective system designed to offer some confidence in decisionmaking rather than an objective design capable of generating a completely confident outcome. Whereas Mill and Mepham have a richer sense of social welfare than Bentham, both are reduced to Hayry's self-admitted consequentialism because they fail to recognize the limits of empirical objectivity.

The closest that risk assessors have gotten to diagnosing their flawed perspective of objectivity is to make the following inquiries: who will benefit from the bovine growth hormone, who will lose from the release of this technology, and, more principally, who will adopt this new controversial technique of modern farming (Barnes & Nowak, 1990; Doyle, 1990)? The assumption that rBST is a technology that is scale neutral, or that it can be equally adopted by all farming regions of the United States and among all dairy producers of different sizes, has been critiqued. Although rBST is a product-enhancing technology with no scientifically proven human health risks, any comprehensive utilitarian assessment must go beyond the simple weighing of possible economic benefits and probable risk costs. For instance, a rational farmer must think beyond mere consequentialism and reasonably question several things: (1) the utility of increased capital and managerial requirements for rBST operators; (2) intensified food and fiber concentrations for treated cows; (3) shifts in production locales of commodities given a substantial decrease in farms due to dairy farmers who will be economically coerced to quit because they have not adopted the new technology; (4) increased labor demands for a biotech farm; and, (5) larger expenditures for veterinary services given known complications with using rBST (Bultena & Lasley, 1990). While this new food biotechnology may result in less reliance on farm uncertainties, significant gains in productivity, and production of the same amount or more milk with a smaller herd, it will have "far reaching, and potentially disruptive, consequences for agribusiness, rural communities, land grant universities and Third world countries" (Bultena & Lasley, 1990, 100).

Currently, there are two major utilitarian approaches to modern agricultural public policymaking: the market model and the social regulation model (Sherlock & Kawar, 1990, 118). The market model emphasizes the role of Mill's rational individual and minimizes government regulation by permitting the use of genetic manipulation of food without substantial regulatory review, provided "that consumers know what they are purchasing through adequate labeling" (Sherlock & Kawar, 1990, 120). The social regulation model ensures the least amount of risk for the greatest number of people by requiring evidence of a food product's safety prior to its marketing. According to the FDA, the regulatory approach is taken because "even a minimum assessment of risk of food additives will require sophisticated physiological and biochemical knowledge that is beyond the reach of most consumers, and poor consumers may find themselves in the market model constrained to purchase the cheapest food products at the highest risk in order to avoid hunger. . . "(Sherlock & Kawar, 1990, 122). As previously referenced, current risk assessments of food biotechnology conservatively attempt to assure the lowest risks to the most people by creating rational standards of safety so that the most untrained consumer can consume large quantities of regulated food, all of the time.

Food products have been regulated under three provisions: the Delaney Clause, the contaminants assessment, and the de minimus risk standard. First, the Delaney Clause automatically prohibits the marketing of any product if there is evidence that a food substance is carcinogenic. Secondly, containment assessments involve using cost-benefit analysis to reach decisions about whether or not to allow marketing food with environmental contaminants. This type of regulation will "consider trade-offs between consumer health risks (costs) and the sum of consumer benefits from having the product (even with its contaminants) and benefits to producers from product sales" (Kuchler, McClelland, & Offutt, 1990, 134). Finally, the de minimus regulatory level of food regulation allows risk assessors some flexibility in trading costs and benefits of their environmental decisions. "The de minimus risk standards do not permit cost-benefit comparisons (as the Environmental Protection Agency may do with pesticides under the Federal Insecticide, Fungicide and Rodenticide Act) but the standards do permit the decision calculus to be more flexible than does the all-or-nothing Delaney Clause criterion" (Kuchler et al., 1990, 135). Recombinant bovine somatotropin, like all animal drugs, whether biotechnological or not, is under de minimus regulation which requires that drug residues in consumable animal products or by-products cause no increase in human health risk.

The central problem with Mill's utilitarian assessment of regulatory policy regarding biotechnology is that it incorrectly assumes that food safety is an objective scientific distinction. Environmental "safety is not an objectively definable and timeless concept" because it involves developing technologies, unpredictable human agents, and a complex environment (Kuchler et al., 1990, 140). Significant risk scholars have argued for some time that biotechnology safety assessment is as much a matter of preferences and choices as it is a science. Thus, universal safety is as poorly constructed as universal rationality. Some risk studies have criticized assessing utility in terms of the fewest risks for the most people, as in the following:

> Turning health and safety management over to a government agency and setting the standards of health and safety is a collective decision. . .When the distribution of benefits from a new technology are not uniform and relatively large groups believe their

> interests are not served, additional collective action attempting to change decision procedures could be expected (Kuchler et al., 1990, 141).

For instance, it has been said that for rBST to achieve a just societal approval, a consensus of safety must be found among consumers, government regulators, and the scientific community.

Social theories of risk have already argued that an ideal democratic consensus will never be achieved because the purely rational method of risk policymaking assumes an objectivity that does not exist in political society. Quite simply, the standard methodology of risk assessments used for determining the food safety of genetic agriculture is problematic because it fails to consider the germane social and economic relations of the policy decision. The technicians, however, do not see a problem because "the empiricist-derived epistemology which has directed most social and natural scientific inquiry for the last three centuries explicitly holds that historical social relations can only distort our. . .abilities to arrive at reliable beliefs"(Harding, 1983, 314-315). Therefore, not only does the preoccupation with objectivity present some difficulties with risk analysis, it also offers some problems for liberal utilitarianism. The methods of empirical observation used in the evaluation of both the practice of risk analysis and the theory of utilitarianism incorrectly presume to produce knowledge that is universally established for the good of all.

Consider Mill's emphasis on achieving individual happiness by means of societal progress. Although there are some instances where the utilitarian maxim is modified, the majority of Mill's politics are justified by their ability to objectively provide the most amount of individual happiness at the least amount of societal cost. Unfortunately, both the theory of Mill and the practice of risk assessment fall short of achieving social excellence because both focus on the seemingly objective principle of utility without considering the subjective notion of preference seeking for the pursuit of progressive excellence. Mill once said that innovators in the ideal society who "shall occupy the posts of honour and power" would be "those who have appropriated most of [the wisdom and] have discovered the secret of human progressiveness"(Mill, 1988, 140). The workings of epistemic communities—either in the science of risk assessment or in the political decisionmaking of utilitarianism—are similar structures of authority that function with a presumption of objective truth when working through the processes to determine acceptable public risk or good. Mill's utilitarianism that produces the conversational credo "it is for the good of all" holds fast to the same presumption of an objective, disinterested, value-neutral process of inquiry that creates the phrase "science has proven." More important, "these ideals are inadequate to guide epistemological debates about contentious issues [, and therefore,. . .] it is deceptive and dangerous to ignore questions about subjectivity in the name of objectivity and value-neutrality"(Code, 1993, 27).

The problem of inserting questions of subjectivity into science and utilitarianism is further complicated by the fact that the "American dream" embodies faith in development of unlimited and unquestioned objectivity. The United States culture has both the Millian desire for vast personal opportunities and the scientific need for flexible social change. In reference to the interconnectedness of objectively derived science and the common good, the following has been said about contemporary American values:

> The urge for material comfort makes Americans ready to accept most gadgets. Pressure has been exerted on individuals and society to search for better methods in technology, including those which give people control over nature and mastery over the environment. As a consequence, high honour and prestige are assigned to the sciences. Since Americans value progress, they tend to emphasize the future rather than the past or the present (Deibert, 1990, 131).

The Millian faith in progress and the American belief that the best is yet to come are frequent rhetorical defences of contested technologies because, despite regrettable risks, technological innovation lights the way to the Holy Grail of progress. Missing from this Millian-like optimism about scientific progress are individuals who have a social and historical context that might be critical of technological progress for its own sake. For instance, individuals who have the experience of being harmed by the "scientifically defined safety" of a food additive have a better understanding of the indirect consequences of progressive technology, not just for human health but environmental integrity also.

With regard to the brave new world of food biotechnology, some of these concerns include:

> ...organisms being created in the laboratory which have never existed in nature before; fewer and fewer farms and corporations controlling most of what is produced, processed, and exported; new biotechnology allowing scientists to consciously shape the genetic destiny of the human race in ways that prove dangerous or destructive; biotechnology accidents with escaping organisms; implications and consequences of university-industry cooperation; property ownership; continuation of eugenics; and social impacts of advances in gene research (Deibert, 1990, 133, numbers omitted).

Notwithstanding the value Americans place on technology, science should exhibit concern for the potential impacts of biotechnology in society that can be reasonably discussed in risk analysis. However, in order to find a subjective perspective within the discipline of science, one must consider science in the nineteenth century sense of the word. That is, when science was a branch of theoretical learning, it was able to critique its own experimental truths. Today, however, the modern understanding of science is an objective investigation of the experimental progress of the material world for human knowledge.

6.FEMINIST CRITIQUES OF MILL'S SCIENCE

Epistemology, the study of the limits to and validity of human knowledge, is an important element in critiquing the American practice of viewing technology without concern for its unwanted consequences. More important, it is through a discussion of the dominant epistemology used in risk assessment, rational empiricism, that many of the ethical implications of utilitarianism are discovered. The history of Western science, including risk analysis, has been about observing, seeing, or depicting the world. This enterprise of observation biases a way of living and learning in the world because it assumes that the human beings doing the research are not interested subjects and therefore do not affect their observable objects, in some cases other human beings. They are comfortable when their methods are metaphorically compared to the faculty of sight. However, some philosophers of science have argued for more empathetic conceptions of scientific knowledge such as listening, hearing, and speaking (Held, 1993). Such an alternative epistemology would view scientific findings as subjective "points of view," "perspectives," or "outlooks" to be considered by policymakers, not as objective determinations about the nature of reality. In this context, the causal explanations of the physical sciences used in risk assessment would be considered just one relevant voice rather than the only valid voice. As such, the political recommendations of its underpinning utilitarianism would be subject to ethical evaluation. Unfortunately, creating an alternative epistemology to replace, augment, or shift the existing paradigm of objective-empiricism may require a systematic redistribution of political power.

Many feminist scholars who work within the field of the philosophy of science have referred to the dominant way of thinking, objective or rational empiricism, as the "male construct of objectivity." One theorist blames this construct for the belief that the truth is found in scientific isolation from personal involvement, political understanding, or social context. She writes:

> The male epistemological stance, which corresponds to the world it creates, is objectivity; the ostensibly uninvolved stance, the view from a distance and from no particular perspective apparently transparent to its reality. It does not comprehend its own perceptivity, does not recognize what it sees as subjective like itself...(MacKinnon, 1982).

Feminist epistemology does not dismiss scientific objectivity, but instead asserts that it is a socialized concept that must acknowledge its subjective components affect both its method and conclusions. It is argued that our standard views of objectivity "are masculine views, and perhaps a feminist standpoint would give us a quite different understanding"(Held, 1985, 299). At the very least, this feminist epistemology offers a different perspective.

"Standpoint epistemology sets the relationship between knowledge and politics at the center of its account in the sense that it tries to provide causal accounts—to explain—the effects that different kinds of politics have on the production of knowledge"(Harding, 1993, 55-56). A feminist standpoint epistemology explicitly admits to being a biased, interested, and engaged theory of human understanding because its purpose is to reveal the social relations inherent in all knowledge and learning (Hartsock, 1998, 107). As such, while gender might be an important source for scientific empiricism or objectivity—there may also be a race and class dimension to the problem—the analysis of a standpoint epistemology is not gender-exclusive. It is only contending that a different voice is needed to understand the false consciousness or destructive socialization of the dominant paradigm (Hirschmann, 1989). Advocates of standpoint epistemology are not interested in replacing the dominant empiricist theory of scientific objectivity with an overtly subjective theory of relativist epistemology; rather, they just want to explain that all knowledge is situated and therefore starting positions of knowledge should be from marginalized voices that challenge the dominant structure (Harding, 1983). From this position, it is argued that knowledge derived from the activities of those at the bottom of the social hierarchies, rather than the top, will be more relevant for the entire community because this knowledge is from a socially determined perspective and, therefore, the results of the research are more apt to be rigorously discussed. Furthermore, thinking from a disadvantaged societal position permits one to question the adequacy of the science and politics of the advantaged in such a way that the epistemology of the advantaged does not allow (Harding, 1993).

Sandra Harding explores the possibility of an exclusively marginalized or feminist science in great detail in *The Science Question in Feminism.* While she ultimately discredits the notion of a specifically feminist science or methodology, she argues that generally women seem to be more subjectively involved with the world and this experience should be incorporated into the mainstream notion of scientific objectivity. "Harding does not believe method is independent of social environment, and so methodological questions cannot be isolated"(Ring, 1991, 33). Harding writes that her "recognition of the fact that science has always been a social product" should not result in one concluding that scientific knowledge is relativist subjectivity, but rather that science is transitional and must be conscious of its political surroundings (Harding, 1986, 137). Perhaps this is best illustrated by the scientific method of criticizing the knowledge of past traditions in order to create new research, different experiments, and alterative observations.

For Harding, all science, including technical risk assessment, is a social science because it is impossible to separate the mind from the world, science from politics, and the internal from the external. Although science is still a method of observation and experiment for Harding, its formulated theories intended to explain experimental results are subjectively limited to accurately predicting future observations. This perspective is a radical critique of the accepted philosophy of science. That is, Carl Hempel argued that science involved formulating hypothetical laws used to create deductive arguments. This was augmented by the falsification theory of Karl Popper, who claimed that no scientific theory could ever be proven to be universally true; rather all scientific explanations are falsified until one theory remains as the correct one. Thomas Kuhn combined some elements of both authors and argued that not all science is mere observations; therefore, the discipline must be separated into the categories of normal knowledge accepted by society and revolutionary knowledge that changes society. Whereas it is easy to discern that Harding would reject Hempel and Popper, one must also note that Harding is not a Kuhnian.

For Thomas Kuhn, science is the pragmatic development of information produced by a community of scholars who develop knowledge through shifts, or in his word, paradigms. While Kuhn debunks the myth of the mad scientist creating empirical truths or indisputable falsities, he still believes that "the truths upon which the scientific community collaborates" are regarded "as pure and objective, immune to influences of society and politics"(Ring, 1991, 37). In fact, Kuhn seems to argue that scientific knowledge is above societal concerns because occasionally science can transform society. This is not the case for Harding, who claims that scientific objectivity is impossible on its traditional male terms. The solution for Harding is to view the truths of traditional objective science as "internalistic" and in constant tension with the "externalistic" or subjective historical, social, and political context in which science is evolving. This evolution thus results in a science that is not universal and, therefore, must be sensitive to its pluralistic truths and democratic validities. Not only should science not divorce itself from its political and social context, but science should engage, reflect, conflict, and learn from the society.

The economic benefits of agricultural biotechnology led the developers of rBST, confident in the universality and objectivity of scientific rationality, to ignore many political variables that influenced the public safety of their product (Browne & Hamm, 1990, 38). "Biotechnology firms, especially because of the importance of this product in setting regulatory precedents for future biotechnologies, should have been as careful in assessing its political feasibility as they were in judging the economy and efficiency of its use"(Browne & Hamm, 1990, 38). Quite simply, despite the rational science of rBST, technological innovators failed to understand the assumptions of rational political science. That is, rationalism involves the calculation of all social, political, and economic values in a public policy, not just those that can be measured by science. The political assumptions of a subjective utilitarianism that allows preference and error are:

> Policymakers must (1) know all the society's value preferences and their relative weights, (2) know all the policy alternatives available, (3) know all the consequences of each policy alternative, (4) calculate the ratio of benefits to costs for each policy alternative and (5) select the most effective policy alternative. This rationality assumes that the value preferences of society as a whole can be known and weighed. It is not enough to know the values of some groups and not others. There must be a complete understanding of all societal values (Dye, 1992, 31).

Because of all of the barriers to rational decisionmaking, some risk scholars have suggested that the advocates of a new biotechnology must have the burden of proof that their innovative technology is safe for all in a political community. That is, a proposed technology must be safe by political consensus, as well as by a scientific one. This is nicely illustrated by risk assessors who were forced by political opponents to identify rBST as rBGH, or recombinant bovine growth hormone, because it would be clearer to the public. Although rBST was a more accurate scientific term, risk assessors failed to realize that their science was out of the laboratory and in the social context of the market.

7.CONCLUSION

Political philosophy's chief enterprise is deliberating about the best social order, but considerations of social order point to an even more basic concern: What constitutes the best kind of life an individual can and should lead? According to many scholars, the roots of political philosophy are found in the words of Socrates and the stories of Plato. In fact, it has been said that all succeeding political theory is simply a footnote to Plato. While that is obviously an indelicate embellishment, it is difficult to imagine any interesting research question regarding politics that successfully would dare to ignore the exhaustive philosophical work of Plato. When one reflects on this chapter's introductory observation by Socrates that societal risks should only be run when they benefit the community, it seems reasonable to inquire about the best type of political structure for assessing societal risks and potential political benefits.

Plato was critical of the accepted Athenian idea of various citizens participating in politics. For him, governing is a distinct art or craft to be completed by an elite ruler or group of elite rulers, trained for such a noble task. That being said, however, it is essential to note that Plato believes that members of the governing class are not simply disciplined rulers; rather they are also philosophers. They, and they alone, understand what is of societal value because they obtain eternal knowledge regarding the transcendental Forms of truth, beauty, goodness, and justice. Furthermore, as a result of their superior intellect, they are reluctant to govern. In order to rule, they need to be forced to leave the private realm of bright philosophical dialogue and enter the dark realm of public discourse. The concluding lessons of Plato's *Republic* are that the feasibility or likelihood of the ideal state is not clear and most societies should live in awareness of the limitation of elites who are not capable of grasping the definitive body of knowledge necessary for ideal ruling. That is, until individuals can be classified by their Golden, Silver, and Bronze souls, decisions regarding the polity should be determined collectively and cautiously.

Expert technicians must realize that environmental risk assessments are not grounded in objective science, but rather are based primarily on a limited view of Bentham's consequentialism. Furthermore, this process of subjective rationality distorts the political liberalism idealized by John Stuart Mill. The "probabilistic assessment of risk," however, can still be useful to liberal policymaking if decisionmakers are mindful of Harding's insight when it comes to trying to live with the smallest risk for the greatest number. According to Harding, good science demands that both the subject and object of discovery be on the "same critical and causal plane," thus allowing its solutions to be reflexive to its problems. Risk issues associated with rBST and agricultural biotechnology prominently demonstrate the defects of using only a rational choice methodology, with internal scientific objectivity, for defining ecological harm. Although the probabilistic assessment of risk is helpful in determining the decision rule, it should not be the only indicator used in accepting or rejecting a risk in American environmental policy. Interpreting risk with only the expected-value approach—methodologically concluding that human and environmental health are anticipated probabilities—is an erroneous assumption for food biotechnology, because scientific safety and policy rationality are not universal political concepts. The utilitarian consequence-evaluation model requires a significant amount of technical expertise, but the ethical acceptability of the technology requires that the political problems be discussed in an open-ended dialogue (P. Thompson, 1998a, 149).

During a significant portion of Western civilization, science was seen as a democratic enterprise. In fact, "there was a time when the philosophy of science inquiry was deeply allied with philosophical arguments for democratic political structures"(P. Thompson, 1998a, 153). Unfortunately, the responsibility for the development of science was placed in the hands of experts and not society, and in their quest for an objective, neutral, scientific methodology, they dismissed all the ethical goals of those concerned with democracy. When this utilitarian point is raised by the strongest bioethicts—Peter Singer, Bernard Rollin, and Tom Regan—they contend that democratic rights must be subordinated to concerns for welfare and utility. That is, in the "General Remarks" to his essay on "Utilitarianism" Mill writes:

> The truths which are ultimately accepted as the first principles of a science are really the last results of metaphysical analysis…and their relation to the science is not that of a foundation to an edifice, but of roots to a tree, which may perform their office equally well through they be never dug down and exposed to light. But though in science the particular truths precede the general principle, contrary might be expected to be the case with practical art, such as morals or legislation...A test of right and wrong must be the maenad, one would think of ascertaining what is right or wrong, as a consequence of having already ascertained it (Mill, 1988, 2).

All too often, philosophy and bioethics are concerned with the isolation and justification of assumptions in order "to identify the principles on which rational moral judgments and decisions must be based" (Maclean, 1993, 5). Indeed, isolated philosophy can be as destructive as isolated science in creating the greatest welfare. In the end, risk assessment must decide whether its means and ends are "the lowest risk for the greatest number," and whether it should be controlled by the technical experts, or "the greatest good for the greatest number" wherein the assessment is an equitable sharing of management and communication between experts and the public. If it is to remain the former, risk assessors must be conscious of the ethical limitations of consequentialism and, if it is to be the latter they must be aware of the subjective influences on scientific objectivity. Interestingly, utilitarianism can be the philosophical foundation for either enterprise.

The male dominated history of risk theory is in many important respects part of the male dominated history of politics. Risk theory, like modern philosophy, has failed and continues to fail to properly address and accept the concerns and experiences of everyone in the political community. In questioning the theoretical and epistemological foundation of political rationalism, feminism can construct new approaches, hypotheses, and concepts about the entire study of politics (Plumwood, 1996). For instance, feminist scholars have revealed the false dichotomy of the public and the private, allowing them to introduce family relations, sexuality, and patriarchical relationships of society as political topics worthy of research.

In the end of this chapter I have noted how epistemologies developed from a feminist standpoint critique the objectivity of science. Most particularly, I have focused on feminists who question the traditional scientific notion of an empirical truth, not because it is a patriarchical truth, but because such an approach divorces the truth from its historical, political, and social context. In the next chapter, I examine the rational practice of risk management from its theoretical foundation of social contract and again use feminism as an analytical tool to critique both the rational practice and liberal theory. Such a descriptive analysis of the opposing positions of environmental risk should aid in the idea that risk analysis might be restructured as a reasonable political process.

CHAPTER FOUR

THE PRACTICE OF RISK MANAGEMENT AND THE THEORY OF SOCIAL CONTRACT

1.INTRODUCTION

Although few if any Americans have ever given their individual consent to eating biotech foods or collectively discussed the relative risks and benefits of their use in modern agriculture, the United States federal government and many private companies are pouring significant resources into transgenic foods and crop biotechnology. Most Americans either do not know about genetically altered food or naively trust that bioengineered products have passed all governmental safety studies and risk analyses (CRS, 1999). In a practical sense, this lack of knowledge exists because the American regulatory system does not require labels on foods with gene-altered ingredients and mandates that biotech companies only consult with governmental agencies while conducting industry studies on safety before marketing the product. In a theoretical sense, the absence of citizen knowledge exists because industry technicians and government food safety agencies powerfully manage the science of risk analysis by using institutions that do not require social contracts or political consent.

In theory, risk analysis is the process of considering the multiple values and ethics that undergird environmental public policy. Irrespective of its liberal intent, foundation, and potential, however, the practice of risk analysis is not open to continued ethical redefinition, value competition, or democratic negotiation. Instead, risk analysis is practiced by experts, and therefore perceived by the public, as a complex scientific examination of how best to administer, regulate, or manage potential societal hazards. In examining the discourse of relevant policy planners, affected segments of the public, and significant political theorists, this chapter explains that many of the difficult questions in American environmental policymaking today are not resolved by scientific consensus but by political compromise. Such an illumination will explain the need for a liberal understanding of administrative risk deliberation and its consequent risk management. As massive environmental risks continue to threaten Americans, policy debates—

albeit technically complex—need to be perceived as subjective or political conversations of cost and benefit, not merely as objective or scientific evaluations of safety and harm. It is important to recognize that rational scientists in the processes of risk analysis consider the political interests of risk because that will allow liberal political theories to inform the scientific process of risk.

Risk analysis must be seen as both a rational technical language for scientists and a reasonable cultural language for society. By describing and analysing the conflict between the rationality of public administration and the performance of liberal political theory in public policy, a meaningful discussion about reform can occur. If public and private administrators of risk clarify the liberal goals of their institutions and agencies, American public policy might be able to better interpret and appropriately react to the risks it actually encounters with respect to the environment. Furthermore, defining the liberal theoretical objectives of risk policy will allow the bureaucratised public to come to terms with the real limits of administrative practices. The rational approach taken in the administrative decisionmaking of risk management is inadequate for analysing all the reasonable risk positions regarding food biotechnology, let alone prescribing complete policy remedies in a democratic state. I therefore examine the contract theory arguments used in the administrative decisionmaking of risk management, which assert that rationally self-interested persons will abandon some personal rights and consent to political powers because doing so is a good decision, compared to the alternative of attempting to manage risk alone (P. Thompson, 1992b).

The most common biotechnological method for genetically engineering food is achieved by splicing a gene from a microbial bacterium and inserting it into a plant seed to confer a certain trait like insect resistance or herbicide tolerance. This is done today with corn, soy, and cotton. Previous commercial applications of agricultural biotechnology include engineered bacterium to increase dairy milk production and to make cheese processes more efficient. More sophisticated gene-splicing techniques are currently being developed to increase production of wheat, fruit, vegetables, fish, and livestock. According to most American governmental regulators, leading businesses, and scientific researchers, there is very little risk and a myriad of benefits in genetically engineering all of these food varieties. Experts argue that biotech foods can resist certain pests, tolerate many chemical herbicides, possess improved processing traits and food yields, withstand adverse soil and weather conditions, and exhibit better ripening, softening, texture, and flavour qualities (FDA, 1998).

Of course, biotechnology is not without its critics. Political opposition has nearly reached the level of a social movement in Europe, and it continues to spread among consumer, environmental, and citizen activist groups in the United States (Gaskell & Bauer, 2001). Although there is no scientific evidence that genetically engineered food is unsafe to eat, some people worry about the potential for unknown allergens, the increase in natural toxic substances, the decrease in nutritional value, and unintended environmental consequences (Burros, 1999b). These environmental concerns have led to everything from protests in which genetically engineered crops were burned to public hearings on the regulatory approval process (Barboza, 1999a). Additionally, some companies—snack food-maker Frito-Lay and baby food-producer Gerber, for instance—have voluntarily rejected gene-altered crops for fear of alienating cautious and distrustful consumers (Barboza, 1999b). Most recently, public scrutiny has occurred over the use of Bt corn, named after the bacillus thuringiensis bacteria inserted into a corn gene to kill the corn borer insect. This public disapproval came about when a variety of Bt corn not approved for human consumption entered the food supply and when university researchers claimed that the pollen of Bt crops kills monarch butterflies.

Risk management is the bureaucratic process of making societal choices about ecological dangers and human health hazards by weighing scientific assessments of harm against social and economic benefits. Within this elite analysis of safety, the notion of a politically "acceptable risk" is developed but never really approved by the public. Therefore, the trust American citizens have in governmental and industry scientists to guard against the risks from modern agriculture—by attending to both the safety of food produced and the integrity of the environment used to produce it—becomes suspect because it lacks any contractual justification. As revealed by public opinion polls and acts of environmental protest against the first commercial food biotechnology, the relationships among citizen consumers, governmental agencies, and industry products are not based on a fully understood trust or politically consented risk (Douthitt & Mayer, 1995; Deana Grobe & Douthitt, 1996; Deanna Grobe & Douthitt, 1999).

The issue of an uninformed and disengaged public is nicely illustrated by a recent survey that asked American consumers whether they supported the use of biotechnology in agriculture and if they approved of genetically engineered or modified foods (Brody, 2000). While there is no substantive difference between the two questions, seventy percent responded yes to the first question while two-thirds were reported as saying no to the second (Brody, 2000). Furthermore, many commentators argue that consumer trust in food biotechnology is waning because the only time they are informed about its use is when there is a problem. This was evidenced by allergy prone consumers who were informed of unlabeled genetically altered ingredients only after the recall of biotech corn products that were not fit for human consumption but were found in many food products and by-products (Kaufman, 2001). It is difficult to create a politics of trust with a community that is so poorly informed and engaged in the topic.

The question that remains, however, is can a liberal democracy respond to the human health and environmental risks that are identifiable to experts and to persons specially briefed, but not to ordinary citizens (Mathews, 1996)? Can a liberal democracy--in theory and practice--respond adequately to the specific ecological challenges of risk associated with new, emerging, and contested agricultural technologies associated with food production? The foundation of political liberalism is the idea that citizens are separate individuals, each with his or her own aims, interests, and conceptions of the good who, therefore, seek a framework of civil rights that will enable them to realize their capacity as free agents, consistent with a similar liberty for all others in the polity (Sandel, 1984). Considering American public policy as an application of this liberal political philosophy is not a radical concept (Ingram & Rathgeb, 1995). In fact, some political scholars maintain that United States public policy could be the ideal academic intersection between liberal political theory and practice (W. Dunn, 1983). In this sense, liberal political theory can be seen as a series of policy alternatives trying to establish the best social structure for democracy.

Environmental risk analyses—such as those involved in the debates over food biotechnology—are incorrectly perceived as technically rational and socially apolitical. Instead, they are based, at least implicitly, on particular liberal political theories. More importantly, risk practices are muddled and risk practitioners are misdirected because they fail to consider the underlying liberal philosophies in their work. Making theoretical foundations explicit, however, should lead to improved practice. Much contemporary research in environmental policy today attempts to solve the policy paradox of involving citizens in complex public policies by simply introducing democratic reforms—citizen advisory committees, negotiated rule making, and multi-stakeholder experiments—in the policy design. Although I agree that such reforms would result in a short-term increase in public deliberation about environmental issues, this type of work fails to discuss adequately the problems with mainstream liberalism in contemporary public policy processes. In contrast, I examine modern philosophies that trade with the political problem of decisionmaking under conditions of uncertainty because of their potential to discern long-term political solutions, such as having a more consent-based citizenship, creating a richer public discourse, and including competing values in decisionmaking.

The organizational function of personal and societal risk in the history of social contract theory is the focus of this chapter. The existence of risk as a political device will first be investigated in the latest enterprise of social contractarianism, John Rawls's *A Theory of Justice*. Like Rawls, industry and policy experts use a self-proclaimed objective understanding of societal risk and, in doing so, fail to recognize the critical importance of citizen-based notions of individual consent and communal deliberation. In order to create a more generous liberal notion of risk management, I turn to John Locke to explore the notion of implicit consent and combine it with the idea of a public discussion of collective interests taken from Thomas Hobbes. Finally, I assert that risk analysis would still be impoverished if it did not include a feminist redefinition of rationality for environmental public policy and public administration. Altogether, this philosophical conversation about risk helps to identify, illustrate, and define the notion of reasonable public consent in the social contract of risk management.

This chapter compares Rawls's philosophical attempt to manage social justice with risk managers who pragmatically attempt to control the technological applications of modern agriculture in the American polity.[15] It is not my intention to jettison Rawls's political theory in risk management, but rather to enhance it with other social contract theories in order to demonstrate the need for a more adequate modern liberalism for contemporary risk analysis. Maintaining a Rawlsian foundation on which to construct a reasonable environmental risk management is important because American public policy can be democratically reformed only if the bureaucratic institutions and its corresponding political citizenry are both liberally grounded. While there has been an enormous amount of criticism directed at Rawls's entire philosophical theory of justice as fairness, there is a dearth of significant work that addresses the marriage of Rawlsian political theory and environmental politics. Although many scholars note the juxtaposition of Rawlsian philosophy and environmental theory, there are only a few works that address the problems of the union of Rawls and environmental policy (Manning, 1981; Pritchard & Robinson, 1981; Singer, 1988; Wenz, 1988). The limited scholarship of Rawls and environmental ethics allows all of the philosophical criticism of the comparisons to be categorized into two camps. These two traditional arguments against Rawls's notion of environmental fairness are important because they are the basis of my criticism of the rationality used in risk management and environmental decisionmaking.

[15] Although John Rawls (1993) is a corrective development of his social liberalism, I assert that risk analysis uses the less satisfactory theoretical devices of the first book to produce just decisions in basic political institutions.

The first area of environmental critique is the central role of self-interest in Rawls's theory. Political liberalism for Rawls is a doctrine whose central goal is that "free and rational persons concerned to further their own interests would accept an initial position of equality" in order to benefit future societal associations (Rawls, 1971b, 11). This reliance on individual self-interest is a major problem for many contemporary environmentalists who assert that it is the same unchecked rational self-interest that produced the anthropocentric economic paradigm that disregards non-human interests and has caused a lot of environmental damage and ecological risk (de-Shalt, 2000, 201). More importantly, because Rawls's theory is "excessively individualistic it neglects the way individual" environmental ethics are formed in "social contexts and pursued through communal attachments"(Kymlicka, 1989, 883). It is essential to note that in order for Rawls's philosophical argument to function, his individuals must be impartial and neutral to their surroundings. This impartial theoretic may create situations of fairness for an individual in the human condition, but such a neutral conception of good has real limits for an environmental ethic that requires preconditioned caring and respect (Barry, 1989; Sinopoli, 1993).

The second category of environmental criticism results from Rawls's focus on political institutions as the locus of public reason. Like Rousseau and Kant, Rawls maintains that significant political problems are to be resolved not in civil associations, but in social institutions. This is problematic for many commentators, for instance:

> Mouffe (1993) has argued that Rawls's liberalism rejects "any possible idea of a common good" and entails the negation of politics (47). Hoing (1993) claims that Rawlsian justice displaces politics with administration and reduces politics to technique (128-61). Connolly (1990) holds that liberalism is tied to a juridical vision of politics (Alejandro, 1996, 1).

These criticisms are largely correct because the implementation of Rawls's liberal politics will require a huge administrative structure of juridical rules to equitably distribute his social product to disadvantaged members of society. Even though Rawls argues that justice should be based on reasoned and informed political agreement, he believes that such a civil society is not located in the public sphere. Rather it is found in the restricted legal sphere, related political institutions, and subsequent public policies (Rawls, 1993).

The primacy Rawls gives to bureaucratic power challenges many environmental thinkers who believe that traditional administrative decisionmaking has not been ecologically rational. More specifically, many environmentalists fear that public deliberation will be lost to the modern maze of interest group liberalism responsible for many of the technological and economic practices that have degraded the environment. Deliberative democrats, discourse theorists, and Tocqueville liberals—all of whom emphasize the importance of inclusive collective debate in developing reasonable decisions to complex environmental problems—have concerns with Rawls's political liberalism, which requires that an elite administer structure, rules, and authority (Charney, 1995; Rawls, 1995). Many environmental theorists contend that ecological solutions to today's complex technological problems require reciprocity, reasonable debate, radical thinking, and untraditional rationalism, all which seem to clash with Rawls's hierarchical legal apparatus and strict adherence to political institutions (Beckerman, 1999; Norton, 1999).

Despite being a theoretical learning process among all affected individuals, risk policy is practiced as an expert rationalization of technical conditions (Wynne, 1992). As a consequence, the administered society assumes that risk is simply the probability of harm measured by experts in environmental policymaking. The foundation of Rawls's social contract theory is that society will be institutionally managed by rational decisions based on risk-benefit calculations or more precisely, based on what representative experts suppose the societal consequences will be. Furthermore, like the risk analyses of food biotechnology, Rawls's game theory fails to incorporate the two most significant elements of modern social contract theory—explicit individual consent and inclusive communal discourse.

In contrast, instead of assuming that all citizens are private, rational, and self-interested, I describe a deliberative process of risk management that bases collective decisions that are agreed to by consent, trust and discussion. This can be achieved when risk management is informed by: (1) Locke's notion of consent so as to include the individual expression necessary for the development of a political community of trust; (2) Hobbes's establishment of a social structure on humanity's discussed common fears and reasons for security; and (3) an inclusive understanding of rationalism that considers and accepts the interests of untraditional political members in society who produce reasonable claims.

2.RAWLS'S GAME OF RISK

John Rawls's *A Theory of Justice* is a radical egalitarian interpretation of liberal democracy that attempts to create a safe and secure society constituted by equitable political institutions and processes (Kukathas & Pettit, 1990). *A Theory of Justice* offers a contractual political philosophy that creates a fair systematic framework of government by refraining from affirming any particular ends. The philosophical enterprise Rawls advocates is a progressive theory of modern liberalism while avoiding some of the weaknesses of utilitarianism, demonstrating a neutral framework of autonomous choice as the basis for ethical principles and defending contemporary social democracy (Lebacqz, 1986). Rawls's liberal theory functions by using a rational choice methodology in a game theoretical design. While the political philosophy of Rawls is quite complicated, its basic contractual theory is rudimentary. Fortunately, the most essential part of Rawls's project for this discussion is his contract method that uses the theoretical devices of the "original position" and "veil of ignorance." To cast away the real aspects of the social world, Rawls designs a condition where an individual can make societal decisions with complete rationality, autonomy, and objectivity (Rawls, 1971b, 516). Although Rawls does not suggest that his rational actors are making particular social policies, he does prescribe that his actors create the foundation, structure, and institutions of society that will create said policies.

The original position and veil of ignorance parallel the technical rationality used in risk management to predict societal risk—both human health and environmental sustainability—independent of context. Rawls's formal model of a society consists of rationally self-interested individuals who are engaged in a bargaining game within an initial situation of fairness called the original position. The original position is an apolitical space for civilized reasoning not unlike the hypothetical starting point called the "state of nature" in earlier social contract theories. The veil of ignorance is a theoretical device that allows individuals to have general understanding of society but no particular knowledge of their advantaged or disadvantaged place in that society. Together these two devices create the ideal rational strategy for constructing an independent place that a single person can use to collectively bargain with the notions of right, dignity, justice, and worth (Rawls, 1992). Rawls writes:

> The merit of contract terminology is that it conveys the idea that principles of justice may be conceived as principles that would be chosen by rational persons, and that in this way conceptions of justice may be explained and justified. The theory of justice is a part, perhaps the most significant part, of the theory of rational choice. Furthermore, principles of justice deal with conflicting claims upon the advantages won by social cooperation; they apply to the relations among several persons or groups (Rawls, 1971b, 16).

It is Rawls's contention that when the benefits and burdens of the distribution of civil rights and societal duties are offered, deliberated upon, and chosen by the parties in the original position, a well-ordered polity will be chosen in a just and fair manner (Rawls, 1971b, 118-119). According to Rawls's bargaining game, once individuals decide to look for the purposes of government, excluding the risks of social circumstances, they are led to the principles of greatest equal liberty, difference, and fair equality of opportunity (Rawls, 1971b). More accurately, Rawls's theory offers a fundamental affirmation of liberty and a limited acceptance of certain inequalities, judged from the perspective of their impact on the position of the least advantaged (Galston, 1980). Rawls explains that once we decide to look for a conception of justice that eliminates the risks of inequity—biological, social, and economic advantage—individuals will choose his secure construction of society.

The provisions in Rawls's contract philosophy that allows autonomous individuals to manage decisions about society with no real knowledge of their own position closely parallels risk analysis that permits public administrators to make societal decisions without an understanding of citizens' real perception of risk in the environment. Rational choice logic, the self-interest based methodology of Rawls, is also the methodology used for testing the risks associated with contested technologies. For example, it might be deduced that if a new beneficial food additive contaminated the food system it would kill 10,000 people. Making a decision on the aforementioned risk calculus is difficult because in real life there will either be a complication with the food additive and people will die or there will be no accident and no deaths.

Most public administrators would feel uncomfortable making such an important decision facing such uncertainty about its consequences. However, in making decisions that determine social policy, risk analysis offers managers the device of expected values of probability, which can function as Rawls's veil of ignorance. That is, risk analysis arithmetically assumes that expected values are like certain values and, therefore, decisions of uncertainty became palatable because reducing the value of a result by its likelihood of occurrence makes consequences seem acceptable. If, using the previous example, it is determined that a new food additive will have only a 1/100,000 chance of contaminating the food system, the new food would then be said to have only an expected risk value of 1/10 of a life. Experts, therefore, erroneously conclude that using rational choice theory and the scientific method—like Rawlsian apologists who use the original position and the veil of ignorance—should result in a consistent acceptance of risk or justice among the public.

Rawls and risk analysis are seen in administrative declarations that seek to control the accidents and unintended consequences of technology by determining a standard of acceptable societal risk for all individuals and the environment. This is illustrated by Rawls when he explains that a just society should maximize not the total but the average utility because it would avoid complicated theoretical arguments in arriving at a public conception of justice (Rawls, 1971b, 161). Similarly, a standard measurement of acceptable risk is necessary, according to policy experts, because the public inaccurately demands greater safety at a greater cost in some areas than in others. For instance, according to public opinion, nuclear reactors must be safer than coal-fired plants and hazardous waste facilities must be safer than municipal landfills (Slovic, 1992). More relevant to this book's concerns, many consumers are criticized for wanting more safety in bioengineered foods than in naturally produced substances. The following paragraph, from a Monsanto researcher, is a good example:

> There is no evidence that recombinant DNA techniques or rDNA-modified organisms pose any unique or unforeseen environmental or health hazards...Greater certainty means greater precision and safety. The subtly altered products on our plates have been through more thorough testing than any conventional food ever has been subjected to. Many of our daily staples would be banned if subjected to the same rigorous standards (McGlouglin, 1999).

Such arguments, however, do not negate the need for a legitimate balance of safety testing and consented risks for judging potential benefits among consumers. In fact, many Americans contend that there are good reasons for concern about a powerful technology that is currently imperfectly regulated.

The rBST case study illustrates an administrative management of risk that is constructed from a probabilistic rational choice, yet results in public resistance, political distrust, and civil disobedience. Many policy experts claim that citizens have acted irrationally because they were advocating much more safety at much greater costs than they demanded in other areas (MacLean, 1986, 28). According to consumer responses before the acceptance of rBST, most citizens surveyed said they wanted milk from rBST-supplemented cows to be so labelled, a significant number indicated they would decrease their consumption of milk if biotechnology was approved, and a majority said their image of milk would change negatively if rBST was involved (Smith & Warland, 1992). There are many instances of negative public perceptions or attitudes toward the risks involved in rBST that diverge sharply from the judgments of the scientists who are in a position to have more accurate factual knowledge of the likely outcomes (Liebhardt, 1993). Public discourse that opposes biotech research and sales is not considered because it is perceived as irrational. This is because it uses reasonable cultural language and examples rather than rational technical language and evidence (R. V. Bartlett & Baber, 2001).

On the surface, the approval of rBST for use in dairy cows seems to be about a controversial technology in the pharmaceutical industry. Underneath, however, is the question about how all biotechnology products are approved in the United States. With the 1992 regulatory system in place, in which biotech companies do all of the testing internally and pay for most of the cost of obtaining a certificate of compliance for a new product, the question of whom the FDA serves in assuring food safety in America comes into question. Although a risk analysis that used Rawls's institutional calculation of accepted risk might have been completed, the average consumer in the supermarket has given very little consent and trust about the safety of biotechnology. Indeed, farmers who administer these biotech drugs into their livestock are not significantly more involved. With big businesses becoming more and more powerful in agricultural decisionmaking, many farmers adopt new ideas of agriculture in order not to be left behind on "the treadmill of technology." At a minimum, the policy of companies not having to label genetically modified organisms must be rethought so that the public can make reasonable choices in the marketplace. At the maximum, the role of governmental testing in maintaining food safety has to be redefined to involve an increasingly cynical public.

The purpose of the contractarian aspect of *A Theory of Justice* is to create a scenario in which the principles of justice are the outcome of rational choice. The original position with its veil of ignorance is, therefore, designed to remove the particulars of human existence that bias choices and affect the fair and just process of discerning the functions of government (Kukathas & Pettit, 1990). The principles chosen will not presume particular concepts of the good, but assume only a minimum about the human condition, specifically that people are free, equal, and rational. The flaw of the Rawlsian bargaining game is that because there are so many uncertain risks that determine social justice, order should not be constructed on the abstract probability that members of society will not be harmed if decisions are made from an "averaged acceptance."[16] While Rawls maintains that the contractarian notion of the original position succeeds in providing the desired neutrality, lay perceptions of risk fail to see that neutrality. Nonetheless, contemporary risk analysis performed today by government and industry continually uses a rationality like Rawls.

It is not that Rawls's rationality in risk decisions is completely out of place; rather, its current administrative practice is incomplete because it excludes all other reasonable concerns as irrational, and therefore irrelevant. In fact, a Rawlsian apologist might argue that because Rawls's theory of justice creates structures and institutions of politics rather than political policies and programs, his administrative and legal organizations could be reformed with environmental participation. However, as mentioned by many

[16] Unfortunately, John Rawls's social contract game theory appears to be more of an attempt to save a method of analysis--rational choice theory--than to demonstrate its explanation of social justice.

environmental critics, Rawls's completely self-interested individual may have problems developing ecological concerns because his neutral being is void of any ethical precommitments. For Rawls, the guiding principle of society is not whether political decisionmaking is ecological, ethical, or even democratic, but whether it is just. And because Rawls's notion of political justice is based on the rational improvement of institutionalised procedures of collective individuals, that privileges the process of just decisions and not the reasonableness of the participants, it functions without needing public discourse (R. V. Bartlett & Baber, 2001). While Rawls's rationalism is based on common sense and established institutions of justice it assumes that its social contract will create a constitution of political stability without democratic reasoning.

The veil of ignorance, intended to remove particular knowledge of experience from its participants while maintaining general knowledge about such things as politics and economics, is destructive for social decisionmaking. As the constitution of risk analysis becomes more institutionalised—like Rawls's principles—its decisions become more about the general process of determining risk and less about the specific people who are pragmatically dealing with risk. Many critics of Rawls's contractualism contend that there can be no such general or abstract knowledge that is not based in practical knowledge (Wolff, 1977, 122). More importantly, the contractarian bargaining game of risk used today is damaging because it incorrectly assumes that all citizens have identical or universally applicable risk perceptions. The only valid construction of a liberal political society results from knowledge and experience that is based on and informed by individual particularities and biases.

Quite simply, lay citizens' supposedly unscientific fears associated with rBST milk are highly relevant to the use of risk assessment techniques in social decisionmaking about food biotechnology. This is probably not the case for Rawls who employs the idea of the original position as his device for understanding the representative concerns of lay citizens. For Rawls, the most just decisions are made from an "isolated realm of politics" by "the general awareness of limited resources and the likelihood of unequal distribution"(R. V. Bartlett & Baber, 2001, 42). Therefore, because Rawls's social contract theory predicts a conception of acceptable risk without actually considering lay perceptions of risks, its practice is an expert definition of societal harm that does not attempt to institutionalise any other reasonable concerns. Rawls's justice as fairness construction is limited because he maintains that answers to ethical questions are not resolved by political debate that could produce a stable consensus of the subject; rather, they are found in the practice of juridical rules embodied in rational social institutions.

The problem of interpreting risk analysis as a rational project and not a political process is further complicated by the administered public's acceptance of elite power in a liberal political society (Stone, 1997). Discerning the proper amount of democratic control, however, becomes troublesome because the authoritative mechanism by which risk is defined and the technological areas in which it is involved are not well understood by the public (Gillroy, 1993). Cases involving environmental risk are inherently technical. It is difficult for lay citizens to analyse issues of food safety, animal well-being, environmental impacts, and social consequences of genetically modified bacteria created to work as a growth hormone for increased dairy milk production. As a result, environmental risk—traditionally seen as a technical judgment of the statistical probability of a hazardous event and its perilous consequences—appears to dismiss political justification, not to need social permission and to eliminate public discourse. Although the scientific language of risk analysis implies that its decisions are objective, apolitical, and rational, many scholars have concluded that risk statements are subjective judgments of differing perceptions of risk (Douglas & Wildavsky, 1982; Mayo, 1997; Whipple, 1992).

Standard risk evaluations that emphasize quantitative techniques fail to address the qualitative causes of public concern about environmental problems (Westra, 1998). Compared to expert opinions, lay assessments of risk reveal what dangers society regards as acceptable and what hazards spur fear, apprehension, and protest (K. S. Shrader-Frechette, 1991). Furthermore, public perceptions of risk not only transcend technical analysis, but are also in some ways more sophisticated because the public has a greater consideration of the values and concerns ignored in the statistical calculations of technical elites (Slovic, 1992). Consequently, the growing focus in risk studies is to persuade the general public to be more rational in its understanding of risk or to convince the technical elite to become more democratic in their evaluation of risk. The problem with both of these solutions is that they neglect the powerful middle position that proposes to achieve some combination of both.

Using modern social contract theories, I discuss the idea of liberal consent and discourse between technical elites and the general public. This prescription for risk analysis is different than other middle positions because it argues that management must become sufficiently more liberal, not necessarily more democratic. Democratically including citizens in policy discussions is not a significant enough political reform to change the distorted modern liberalism used by environmental management teams in current political administrations (Paehlke, 1994). A just environmental administrative decisionmaking process would function by deliberating and communicating with many different citizens (Paehlke, 1990). This is best achieved by the standard of modern liberalism: contractual consent. It is the notion based on the premise that citizens should civilly live together and that accepted social expectations about power will be exercised for solving disputes about personal interests and political obligations (P. Thompson, 1992a). Having already identified the rational practices of the expert knowledge of risk, it is now important to explain the liberal theories underlying the public perception of risk in order to discuss any possible connections over the current divide. Such an enterprise might create a meaningful discussion of ecological democracy or a reasonable understanding of environmental liberalism.

3.LOCKE'S LIMITED GOVERNMENT OF RISK

John Locke's social contract introduced in the *Second Treatise of Government* joined together what seemed to be conflicting notions: legislative superiority, majority rule, consent of the governed, and order by rule of law (Wiser, 1983). His notion of classical liberalism has probably been the most influential philosophy of democratic constitutionalism. For instance, Locke's ideal political system—organized by liberty, but preserved by the separation of governmental powers—is the foundation for the American structure of politics. Most important for this conversation, Locke stressed consent as the basis of political power and suggested restraints on the power of the majority to prevent violation of laws of nature, property privileges, inalienable rights of citizens, or anything in the fundamental contract (Curtis, 1961a). Furthermore, as a consequence of Locke's claim that all are born with a title to freedom and that all legitimate authority is founded in consent, a political society of trust is formed.

For Locke, government was created to produce certain benefits by having individuals agree to risk certain natural roles, relationships, and rules in society. Locke writes, "...there and there only is political society where every one of the members have quitted this natural power, resigned it up into the hands of the community in all cases that exclude him not from appealing for protection to the law that established it"(Locke, 1993, 304). The social contract is agreed upon for safe and peaceful living and in order to secure enjoyment of property. It is a contract made among equals to create a society, and then made between that community and the ruler. Power comes from the people, and law is rooted in common consent. Locke famously writes, "Thus we may see how probable it is that people were naturally free, and by their consent either submitted to the government of their father, or united together out of different families to make a government. . ."(Locke, 1993, 319). Political authority—the power of the whole community—was found in the will of society as expressed in the sanction of the majority and performed by agents or representatives of the government.

Central to Locke's political philosophy is his epistemology. Human understanding for Locke was determined by reason and toleration. Knowledge consists of seeing relations among ideas derived from sense experience and as a consequence of the "limitation of human understanding, one could talk only about probability, not about certainty" (Curtis, 1961a, 326). More importantly, because communities do not function under the collective notion of certainty, one of the central political requisites for Locke is trust. Many scholars have argued that Locke asserts that tolerance is a precondition for a politics of trust. In his social contract, therefore, Locke demands that individuals must arrive at some moral criteria for determining when to trust peers, rulers, and institutions (J. Dunn, 1985). Locke writes, "And thus every man, by consenting with others to make one body politic under one government, puts himself under an obligation to everyone of that society to submit to the determination of the majority, and to be concluded by it (Locke, 1993, 310)...and the love and want of society, no sooner brought any number of men together, but they presently united and incorporated, if they designed to continue together"(Locke, 1993, 312). For it has been famously said: Locke does not question whether or not we are going to have a collective life, he is concerned with what the quality of that collective life will be.

Integral to the collective contract which uses necessity, consent, and trust is the voluntary self-commitment to the majority, commonwealth, and sovereign. This is very different than the Rawlsian contract that probabilistically determines acceptance by abandoning notions of trust, agency, and consent. Recall that Lockean political trust is grounded on the affirmation that the risks of complying with authority or others are acceptable. Nonetheless, it requires that participants in the political practice and occupants of institutional roles are agents working with the consent of the people. While it would be absurd to expect risk managers to require the consent of every individual in the political society, it is unjust for the representative government to make decisions that risk the body politic without informing it. Locke argues that living under the privileges and protection of the state can only be just and truthful "by positive engagement, and express[ed] promise and compact"(Locke, 1993, 324). This general Lockean social contract, however, does not appear in some contemporary cases of risk theory and practice.

Food safety regulations have the potential to preserve Lockean citizen sovereignty by informing consumers of risks that would be difficult or impossible to detect without special training. In risk assessments involving food biotechnology it has been said:

> Though few people have given explicit thought to the ethical principles that govern food choice, consumer sovereignty provides a reasonable interpretation for the implicit ethics in traditional norms for buying, selling and consuming food. This principle presumes that food consumers give informed consent to the purchase or consumption of food in normal market transactions, and food suppliers have discharged the bulk of their ethical responsibilities by providing information. Ingredient labels inform consumers of the constituents in processed foods, permitting consumers to choose or avoid foods based on health, aesthetic, religious or even purely idiosyncratic considerations (P. Thompson, 1997, 34-38).

This contract is, of course, contingent upon the consumer having the "right to know" what the government demands and, therefore, what the industry producers are free to do. It is easy to observe the importance of and emphasis on scientific probabilities in the analysis of risk in food safety (Feder, 2000). The correlations between exposure and harm are extremely important in setting policy for food safety and quality, but they are not the only criteria for risk policy. Scientists who talk of risk from biotechnology only in terms of probability dismiss the public's opportunity to raise reasonable concerns about agency and responsibility (P. Thompson, 1998b).

The reasonable consumer is completely divorced from the governmental agencies of the (FDA) and the Food Safety Inspection Service (FSIS), which regulate based on a scientific or probabilistic interpretation of safety (P. Thompson, 1998b). Industry and government do not need any individual consent for their power in risk management. The best example of this is seen in modern biotechnology that has the capacity to introduce new constituents by genetically engineering whole foods such as milk. According to current American policy, the government does not have to inform consumers that plants and animals have been transformed through recombinant DNA technology or that microorganisms were scientifically transformed during the production process. This is because properly conducted gene transfers pose considerably less risk to consumers than conventional techniques for developing or modifying foods (FederalRegister, 1992). Nevertheless, consumers who find food biotechnology ethically or environmentally questionable have almost no capacity to avoid recombinantly transformed food because current policy does not oblige producers to inform consumers (P. Thompson, 1997).

The government's regulations regarding what food labels must disclose are not based on consumers' desire to know. For instance, a food does not have to be given a common or usual name nor does the manufacturer need to disclose information about how the material is produced or to indicate any consequences that may arise from the use of the product (FederalRegister, 1993).

In a modern democracy, this policy is alarming because citizens should have the "right to know" and, despite scientific consensus on the safety of certain technologies in the past, other technologies have later been found to cause considerable health and environmental risks. In fact, some caution about food biotechnology recently has been suggested because government documents show that agency scientists disagree on altered food. According to internal government memoranda read at public hearings, several FDA officials have disagreed with the agency's conclusion that genetically engineered foods can be regulated in the same way as conventional food varieties (Burros, 1999a). As a consequence of the absence of Lockean consent in food risk assessments, industry and government perform most risk analyses. "Their analyses include three main stages: (1) identification of some public or societal hazard; (2) estimation of the level and extent of potential harm associated with it; and (3) evaluation of the acceptability of the danger, relative to other hazards"(K. S. Shrader-Frechette, 1991, 5). Once the experts have completed the aforementioned tasks, policymakers then determine the best way to manage the risk of the particular threat. This is commonly done through regulation, moratorium, prohibition, or taxation.

For instance, rBST is regulated in the same way as natural milk and, therefore, the government does not require that it be labelled or studied for long-term societal effects. In fact, a United States federal appeals court judged that the free speech rights of dairy producers legally outweigh consumers' right to know what is in their milk when it suspended a state law requiring labels on products from cows given the bovine growth hormone (StaffReporter, 1986). In dissent, one circuit court judge wrote that consumer interests in the rBST effect on human health, well-being of cows, and economic impact of rBST milk production were legitimate, because the real concern of dairy producers is "concealment"(LawReporter, 1996). The appellate court, nonetheless, found the state's interest in disseminating information to consumers was negligible because it violated the International Dairy Food Manufactures constitutional right not to speak, demonstrated nothing more than consumer curiosity, and contradicted the FDA's conclusion that rBST poses no threat to human health (LawReporter, 1996).

The majority of the citizens in the United States seem to trust government and industry scientists to guard the safety of their food and the integrity of the environment. As previously mentioned, however, that relationship between consumer, government, and industry is currently based on a disputed power and not respected trust. The scientific institutions of government and the organizations of industry are at the centre of a powerful and technocratic Leviathan. "They control the definitions that are used to translate regulatory mandates into operational terms," run administrative departments of safety, rarely attempt to communicate technical concepts to the public, and fight to "maintain the structure of the Leviathan, for it assures their status and the continuing demand for their services"(P. Thompson, 1998b, 236). Governmental food safety agencies and other scientific regulators of biotechnology operate without implicit individual consent in the current regulatory process.

Monsanto, the only company approved to sell rBST—under the trade name Posilac—opposes legislation to require food biotechnology labelling and refuses to voluntarily disclose this information on its consumer by-products (FDA, 1999b). The drug company maintains that any label would have a negative impact on sales because it would stigmatise a safe and beneficial technology by causing consumers to question the human safety and social consequences of rBST simply because the product is unfamiliar. A government label, one that certifies the presence of any recombinant DNA technology in food, would help assure individual consent but would not completely satisfy a contractual liberalism. Granted, if familiar natural milk were offered as an alternative to labelled rBST milk, it would provide some implicit consumer consent. Although labelling would aid in consumer sovereignty, it would fall short of assuring political consent because it would offer no democratic space for informed consent to emerge in a community discourse.

The implicit consent that a consumer gives in purchasing a product is not the same as the explicit consent that a citizen gives when participating in a political decision (Pateman, 1970). More importantly, it is philosophically dangerous to consider the aggregate principles of political interests, opinions, and decisions as though they were all rational market values. Quite simply, "some values require different procedures, not just consistent outcomes, and . . .it is not reasonable to treat all social decisions consistently in this" rational manner (MacLean, 1986, 24). Thus, it is practical to conclude that public fear and opposition to biotechnologically altered milk has been misperceived because policy experts have focused their advocacy of rBST exclusively on its safety. That is, the source of fear and opposition might not really be about the decision to approve use of rBST, but about the way that decision to accept rBST was made. At the very least, it reflected the impoverished modern liberalism of administrative decisionmaking and the undemocratic practice of environmental risk management in food biotechnology. The perverted social contract of risk management that rationally defines power is similar to the distorted liberalism that the scientific-rational community uses in the utilitarian assessment of risk—where the decision rule is the "lowest risk for the greatest number" instead of the "greatest good for the greatest number." Here, however, the social contract split is between scientists who define modern liberal power as the institutional control of safety and social advocates of modern liberalism who define power as the public acceptance or legitimisation of institutions by consent or trust.

Many political scientists who study environmental policy have argued that the factors that go into environmental decisionmaking include scientific knowledge, specific technology, complicated economics, and political values (Dryzek, 1997; Kraft, 1996; Paehlke, 1990). Given the technically specific knowledge of environmental policy today, however, many liberal values are often subordinated to technology, science, and economics. As a result, risk analysis is undemocratic because bureaucratic, scientific, and industry experts control the process of developing environmental policy. Unfortunately, this expert control is evident not only in the lack of individual consent in food biotechnology, but also in the absence of public discussion of consequences. The irony is that social contract theory—based upon principles of just order formulated by a collective agreement of people for a private sphere within which they can partially determine their own destiny while being shielded from external threats—is the underlying argument in risk management performed in administrative decisionmaking. Despite communal acceptance being present in Locke's contract, the notion of rationally calculating social risk is better emphasized in Thomas Hobbes's political theory. Regardless of not always being categorized as a liberal thinker, Hobbes's contract must be used to investigate if administrative decisionmaking allows collections of individuals to accept expert estimates of social safety by asking: Does risk management presuppose or provide a model of communal consent?

4.HOBBES'S RISK CALCULUS

Hobbes is the ideological advocate of realism, the belief that sheer political might makes social order. Hobbes is the theorist of political power who defends the view that the only alternative to anarchy is absolute subjection, and that an overriding fear of death drives people to embrace a sufficient, but necessary, authority. His major work, *Leviathan*, is a philosophical rationalization for the creation and preservation of an authority to provide maximum liberty and ensure political stability. Hobbes's central concern is to diffuse the hazardous effects of living without order. His relevance to this project is found in his notions of the state of nature, communal contract, and sovereign power. With these Hobbesian devices, the modern theories of a public security and discursive community are properly developed to justify the contemporary practice of risk management. Furthermore, in looking at the traditional critiques of Hobbes, the final elements of a liberal risk regime are completed.

Hobbes argues that in the state of nature every human has a natural right to do whatever he or she thinks necessary for preservation, including the destruction of another human. Since the powers of individuals are simply natural rights and essentially equal, all power is realized in competition with others. Human beings escape the intolerable risky state of nature by giving up their natural right to defend themselves, making a social contract, and creating a sovereign. The sovereign becomes the sole authority of the rights of individuals and secures peace and order in the public sphere of civil society. By Hobbesian metaphor, political society is a watch, designed and operating for a practical function. As the watch has mechanisms, behind the political society are human needs. In order to describe the human mechanism of need, Hobbes creates an original condition called "the state of nature."

According to Hobbes's practical observations, humans are motivated by just two things: seeking pleasures and avoiding pains. From this perspective, Hobbes concludes that the core of all political justice is human desire, both appetite and aversion. Power is defined as the means to satisfy these desires. The equal powers of individuals are therefore realized in "whatsoever therefore is consequent to the time, wherein men live without security, that what their own strength and their own invention shall furnish them withal"(Hobbes, 1994, 76). To eliminate collective fear and the state of unbearable risk, humans create a social contract through a dialogue about the benefits of a secured society and the costs of an all-powerful sovereign state. It is agreed that although the sovereign power is significant, it is better than the risks of the original state of nature. Furthermore, subjects can resist the sovereign if their lives are at risk, for that would be an indication that the sovereign is not performing the function for which it exists, providing security. That is, "the obligation of the subjects to the sovereign is understood to last as long, and no longer, than the power by which he is able to protect them"(Hobbes, 1994, 144).

The political philosophy of Thomas Hobbes is important within the history of political theory because the possession of power by the sovereign is justified not by appeal to tradition, divine right, or heredity, but in a utilitarian fashion (Warrender, 1957). Society and the all-powerful state are deliberate human creations, the product of the unity of individual wills. It is essential to note, however, that Hobbes's society is an artificial convenience created by the fears of the individuals, and it seeks only to preserve the individuals' interests and accommodate their singular needs (Wiser, 1983, 194).

While Hobbes creates a liberal order, it is built on the assumption that the undesirable consequences of the civil state are less certain to occur than the undesirable consequences in the original state of nature. Hobbes argues, "that universal egoism, which is merely welfarism at the individual level, could be channeled by strong government to produce universal welfare and that egoists, for their own benefit, would want such government"(Hardin, 1991, 156). Hobbes's political order is therefore a utilitarian contract of fear.

The philosophy behind risk management is based on the same calculations that underpin Hobbes's social contract theory. Risk management is simply a realistic device of social control. That is, the history of American environmental policymaking with regard to health and safety has been a centralized, command and control model of politics. Although the causes of fear have changed since Hobbes's project, the uncertainties and limitations of freedom today, according to risk management, have the same effect as the *Leviathan* does. That is to say, risk analysis suggests that the basic operation of democracy cannot adequately interpret and control the dangers of a technological society. In fact, many argue that such an interpretation is impossible because American culture is based on societal progress driven by industry, science, and technology without environmental concerns. Thus, rational experts must control it.

According to Hobbesian liberal philosophy, however, a collective can resolve dominant power relations with a broader pattern of discourse—through an inclusive and consensus based liberalism, the current undemocratic power of environmental management can be changed. In fact, Hobbes's political realism simply demonstrates that knowledge claims are the desired product and unintended consequence of the relations of power. Rather than stating, like post-modernism, that there is no choice but limited politics, Hobbes seems to argue that modernity has inappropriately chosen limited politics. Thus, the only way to reject or consent to this control is to incorporate a public space for risk discussions, hitherto lacking in the scientific assessment of risk.

Environmental scholars have just begun relating ecology and democracy. This literature discusses the following: "initiating systematic changes via overhauling political structures and procedures; assessing the merits of political centralization versus decentralization; and linking environment with other social conditions like communitarianism and egalitarianism"(Zietler, 2000).[17] According to many theorists, the advantages of "civic environmentalism" or participatory environmental policy and administration

[17] Generally, see Zietler (2000) and specifically, note her introduction to chapter two, "Democratic Theory and Environmental Politics," 16.

are flexibility, accountability and legitimacy (Mathews, 1996; Press, 1994; VonSchomberg, 1993). This type of liberal environmentalism is essential for risk analysis because the loss of something valuable in order to gain something of greater value needs to be open to changes in objective data and subjective perceptions (Byrnes, 1993). Recognizing that a voluntary risk is more acceptable than a coerced risk is central to understanding the public outrage against unknown or coerced risks. Consider, for instance, the voluntarily accepted risks of smoking cigarettes compared to the coerced risk of eating genetically modified food not approved or licensed for human consumption. The former is more accepted than the latter because it is personally consented to and publicly debated.

It should be noted that Hobbes's collective liberalism changed the virtuous justification of political power (Drury, 1988). Recall that for Hobbes, society is formed in order to avoid certain consequences. In particular, human beings joined society to avoid a premature and painful death. As a result, Hobbes's society is formed by the threat of the greatest evil rather than by the traditional promise of the greatest good (Strauss, 1959). The decisive feature of Hobbes's universe is "the exclusion of the traditional notion either that human beings might naturally be joined in seeking a common human excellence or that there might be, for any man, a conclusion to his activity which he might seek and recognize as the common good"(Oakeshott, 1962, 355). Organizing power in order to create peace by fear is another problem with risk analysis. Whereas I have deference for Hobbes's contractual consent formed by public discourse, I would like to diminish Hobbes's uniting principle of fear.

The traditional ethics of politics is lost for risk theory because, like Hobbes, it has a disregard for the idea of a general good. Just as Hobbes constructs an absolute state from its atomistic parts, risk managers control society through its individual parts. For both, the notion of politics as a virtue for the entire community is lost. Good is not simply the object of one's desires, as Hobbes claims, nor is it the expected value of safety, as the technical risk community claims. In order to develop a more comprehensive contractual risk management, the political power to secure peace in a state must originate from an ethical discourse or communicative political space. Whereas realist political power is a logical theory and systematic practice, it incorrectly denies the idealism of a tradition that views the noble and the just, independent of the pleasant and the preferable (Drury, 1988, 134). Similarly, risk analyses conducted only by the technical risk community are very efficient but usually inequitable in practice.

One of the greatest lessons of the rBST case in the United States—one that has not yet been learned in the American regulatory process of food biotechnology—is that the absence of any political forum for debating the social consequences of a contested biotechnology is undesirable (P. Thompson, 1998b, 43). That is, the risk assessments completed by the industrial and governmental Leviathan do not accurately consider the different concerns of all citizens. Recall that the genetic engineering that made rBST possible was performed on a microbe, which was produced in an animal drug and then finally inserted into a dairy cow. Because so many people had no place to discuss their ethical concerns the controversy surrounding rBST became the first real case study of the unintended social consequences of food biotechnology.

In a liberal construction of democratic life, significant policy decisions that greatly affect society, like the production and sale of food, need to be debated by the public. A public domain is a discursive space in which a political community is able to develop common interests and forge common purposes through participatory dialogue (Barns, 1996). Although multiple governmental agencies investigated the risks associated with rBST, none attempted to enrich the discourse by extending its boundaries from expert rationalizations into public life. Granted, the government has probably "adequately addressed all the human food safety, animal safety, and drug efficacy points pertinent to the new animal drug application"(GAO, 1992).[18] But the relevant characteristics that contribute to the public perception of risk in the rBST controversy exceed the "acceptable risk" calculated by the technical community. The consequence of a rationally determined acceptable risk, according to one theorist, is the notion of public outrage. This embedded nature of risk is explained as follows:

> Many experts resist the pressure to consider outrage in making risk management decisions; they insist that 'the data' alone, not the 'irrational' public, should determine policy. But we have two decades of data indicating that voluntariness, control and fairness, and the rest are important components of our society's definition of risk. When a manager continues to ignore these factors—and continues to be surprised by the public's response of outrage—it is worth asking just whose behavior is irrational (Byrnes, 1993, 337)?

The United States government offers incomplete risk assessments because it continually regards all public concerns with rBST as irrational and, thus, not worthy of engagement. However, something could be worthy of political concern and still be irrational.

For example, in an unprecedented event the EPA sponsored some multi-faceted studies to answer questions about Bt corn pollen and the monarch butterfly (ARS-USDA). Not surprisingly, the scientific evidence showed that the risks monarch caterpillars have due to genetically modified plants is

[18] Final (GAO) Report for Congressional Requesters is dated August 1992.

negligible. The event is noteworthy because the scientific investigation was a response to a public outcry of irrational concerns. That is, Bt corn had been registered since 1995, all studies regarding non-target species had been reviewed, and the crop was successfully being regulated by the 1986 coordinated framework of product familiarity. Nonetheless, the EPA was able to allow science to guide decisionmaking because it considered the public outcry a reasonable claim.

As a consequence of not having a public space to discuss citizen objections to the contested technology, reasonable critics made many unreasonable claims about rBST. For instance, the economic effects of rBST on the dairy industry were not as dramatic as originally projected by its opponents. Governmental accountants concluded that reforms underway in the industry were "reinforced but not significantly changed with rBST adoption; and industry effects of rBST ultimately depend on the flexibility of the dairy price support program"(ERS, 1990).[19] It does not matter, however, if accusations that rBST is risky or unsafe were warranted, truthful, or suitable. They were and are completely understandable because there is no amicable place to discuss the problems people have with rBST in the family, workplace, and home. This is illustrated by recent claims—despite no scientific evidence linking the use of rBST to human health threats—correlating the use of recombinant bovine growth hormone to mad cow disease and its human variant, Creutzfeldt-Jakob disease (Westra, 1998, 129).

Most of the resistance to rBST is a result of people being unable to raise reasonable objections for serious political concerns. That is, the political acceptance of rBST in 1994 did not even meet the impoverished standards of communal consent of fear described by Hobbes in 1651. Interestingly, if the process of approving rBST had incorporated a public space for reasonable debate, the FDA would probably have still approved the contested technology, but there would have been less political opposition to rBST, and the public confusion apparent today over emerging biotechnologies would be reduced. My basis for this conclusion is that since the bovine growth hormone was approved, the following has occurred: (1) About 30 percent of the nation's dairy cows have been injected with the gene-altered drug; (2) sales of organic milk represent 0.3 percent of the 75 billion dollar milk industry in this country; (3) consumption of milk since 1994 has increased; and (4) rBST is considered a viable political topic in American agriculture policy in small dairy communities and international treaties (Burros, 1999a, 134; Powell & Leiss, 1997). It is important to note that although industry developers of new foods typically spend years and conduct dozens of internal studies to evaluate whether products like rBST milk are commercially safe, this

[19] Please see the final Economic Research Service (ERS) report dated June 1990.

process does nothing to inspire social trust and acceptance. In fact, one could argue that a process that does not use independent safety testing combined with public information sharing will never eliminate the anxiety and suspicion that the public has about food biotechnology.

The public policy of decisionmaking through risk evaluation—a political practice without virtuous standards of measurement—is limited, complicating the management and use of risk assessments. The limitations do not eliminate the need to provide credible risk analysis in order to control the environmental problems of today. The limitations do, however, demand a philosophical practice capable of considering more public values in environmental decisionmaking. A Hobbesian risk analysis offers no principles for developing a more substantive vision of the common good. The following questions, however, arise when attempting to incorporate participation in a contractual liberalism:

> The first concerns the inclusion of "difference." How are we to encompass within a democratic process oriented towards the forging of some substantive "common good" the diversity of practices, beliefs and voices that constitute our increasingly pluralistic society? Second, given the recognition that democratic participation is also a moral practice, what kinds of selves do we need to be in order to be committed to a particular vision of the good life and yet at the same time live in a peaceful harmony with those who are different? Third, how can we enable productive and peaceful dialogue between communities holding competing and often incommensurable ideological visions (Barns, 1996, 107)?

In order to create a consensual, inclusive, and discursive risk analysis, two things must occur. First, the definition of rationality must be expanded to include political considerations, and second, non-traditional members of society must have some opportunity to participate. Such a reform of American environmentalism would develop a risk analysis that works on the basis of reasonableness of the common good and not from the rationality of the general fear. This is conceivable if risk management can integrate the realist control of the technical community while valuing the social positions of risk from the public community.

5.RISK PRACTICE AND CRITICS OF SOCIAL CONTRACT THEORY

As argued, the prescription for a comprehensive risk management would entail a social contract that constructs a community that openly discusses its societal risks, human health safety, and environmental standards. This creates a dilemma because, having already articulated the primacy of explicit individual consent within a communal society, I need a complex liberal politics characterized by the advancement of the security of the society, not simply the democratic principle. That is, while I am dissatisfied with the impoverished social contract of contemporary risk management, I fear the extension of political decisions and the acceptance of the will of the majority as the only valid political wisdom in multifaceted environmental problems. Therefore, what is needed is a risk management that allows engaged citizens to obtain a sense of purpose by freely practicing their choices in a political community of reasonable security.

In a democratic political community, citizenship carries the promise of self-government. Fundamental to an enduring self-government is active political participation among the democratic citizenry. The traditional American political science of pluralism—egalitarian voting, majority rule, and electoral representation—suggests that those who participate in the political process are frequently listened to in a democracy, while nonparticipants are often ignored (Mansbridge, 1980). Many contemporary political theorists contend that in the current American "politics of expertise," however, the proposition of self-government is dismissed by both the mainstream policy stakeholders and the administered public citizenry (Fisher, 1990; Ingram & Rathgeb, 1995; Mathews, 1996). Since the writings of Max Weber, who stressed social relationships of hierarchy rather than equality, the study of democratic public policy has been wrestling with the problem and power of administrative rationality. Administrative rationalism is the problem-solving discourse that emphasizes the role of the policy expert, rather than the citizen, in public policymaking (Dryzek, 1997).

Within the arena of environmental bureaucracy, the notion of an ecological rationality has been developed to redirect administrative rationality (R. Bartlett, 1986). Ecological rationality incorporates ecocentric values for the long-term implications of environmental policy into the anthropocentric short-term decisionmaking process (R. Bartlett, 1998). Operating within this scholarship, my research attempts to make the American bureaucratic culture more liberal, critical, and effective. If local culture can demystify expert policymaking and if the technical community can appreciate and respect the logic of local citizenry toward environmental concerns, points of intersection will be possible in American society (Whipple, 1992). In dealing with the problem of administrative logic in risk analysis, risk public policy must learn something significant from feminist political philosophy.

Feminist theorists question, for instance, the masculine values of power relationships that dominate many ideas in political theory. It is often concluded that such relationships must be reconstructed in order to include all members of the polity. In changing the rationality of risk management by accepting feminist political theory, risk management embraces inclusion of more people into the contract. Feminists interested in democratic processes generally maintain that many of the great philosophers of Western Civilization were men who exaggerated their isolation from women and children as they evolved their ideas about happiness, love, citizenship, government, and virtue. As a result, their philosophical legacy, created without women's participation, has tried to explain the world by erroneously concluding that women were not members of it. Specifically, some feminist authors contend that many of our central concepts such as freedom, individualism, contract, and justice rest upon arguments issued from a collectively isolated male world that is not representative. For this discussion, the liberal ideas of Hobbes, Locke, and Rawls need to be reinterpreted with a more inclusive theory.

If liberalism rests on the assumption that "individuals" possess their own bodies and freely engage in social contracts in the political realm, the individual is simply male because it was man who possessed his body in traditional contract theory. As a result of masculine sexual mastery involving possession and access to sexual property, women are subordinate to males in this contract. Hence, the concepts developed to explain traditional political relationships are inherently masculine because women lack the civil standing given to men in traditional contractual liberalism. Some feminists argue that the idea of social contract itself is male domination, and its use as a concept by feminists shifts the discourse within a patriarchical paradigm: "The feminist dream is continuously subverted by entanglement with contract" (Pateman, 1988, 188). This radical critique concludes that a free order cannot be a contractual order and that "the sexual contract and the social contract, the 'individual' and the state, stand and fall together" (Pateman, 1988, 232).

The controversial argument of this type of feminism is that political obligation constitutes an unsolvable problem in the liberal democratic state. And the problem can only be solved within the context of women being considered participatory members. Unlike other critiques of social contracts, a feminist analysis offers a mixture of condemnation and gratitude to the notion of liberalism. This type of scholarship adequately explains a policy framework that had positive and destructive effects on the effort of developing knowledge about the human condition. Here feminist political theory is used in order to rerationalize risk management by expanding the political community allowed to debate the risks and benefits of agricultural technology. With regard to liberating risk studies, feminist scholarship has recommended the following re-evaluations: "(1) how risks and hazards are identified and assessed; (2) broadening the public discourse on uneven or disproportional burdens; (3) politicizing action in influencing acceptability judgments; and (4) outright resistance to new technology and existing risks and hazards"(Cutter, 1993, 186).

Most feminist work on rBST discusses the quickly mobilized grass-roots opposition to the first genetically engineered food. As a consequence of pregnant women and children being the largest cohort of consumers, women's groups were able to stage significant boycotts of biotech milk—regardless of FDA approval and convincing evidence regarding the human safety of the product—simply because they prefer milk to be "natural"(Cutter, 1993). The significance of a feminist perspective on risk management, however, is that because of the historically marginalized role of women in social contracts, their voices are ideal for questioning an administrative management that makes public decisions on risk after listening only to voices of policy rationality and scientific probability. In extending the polis to consider the individual consent of nontraditional members and involve them in a valid political conversation about controlling risk, the informed judgments of experts may actually consider or rerationalize the reasonable comments of the "uninformed" public.

Rerationalizing means reconceptualizing, reforming, or redefining the notion of rationality to include reasonable political debate from nontraditional members, who were excluded on the basis of their irrationality. It is essential to note that this philosophical move is arguably not in correspondence with most ecofeminist scholarship on biotechnology. For instance, some authors in *Biodiversity: A Feminist Reader on Biotechnology* decry genetic engineering as a patriarchical science that systematically destroyed indigenous women's farming, and others claim that biotechnology should be discontinued because it disproportionately positions women and children at risk. Instead of making a substantive policy statement, my interest is in delineating a possible reform to the policymaking process. As seen in this analysis, my theoretical argument with food biotechnology is not so much with the product of a current policy, the acceptance of rBST in this case, but with the process of policymaking. What my analysis shares with these feminist critiques is the notion of decisionmaking that emphasizes relationships rather than contractual approaches that stress the role of individuals apart from their social networks. Relationships are more significant for Hobbes and Locke than they are for Rawls. However, none of these authors make them essential, as they need to be for a reasonable risk analysis.

The notion of a social contract with a relationship ethic can be seen in environmental concepts of stewardship, sustainability, and agrarianism (P. Thompson, 1998b). Furthermore, this relationship ethic is important because of its focus on citizen reasonableness rather than technical rationality in rerationalizing risk analysis. Rerationalization pursues ecological rationality in terms of reform of the bureaucratic, market, and legal culture of society (Murphy, 1994). While still using traditional rationality, it considers nontraditional arguments in constructing decisions. For risk management, this would mean including lay perceptions of risk in probabilistic determinations of risk. In terms of feminism, this is achieved by changing the social hierarchies of decisionmaking by introducing the importance of political consent through communal discourse and relationships. For the case study, this would mean that Monsanto and the Federal government would consult with environmental, consumer, and other groups with concerns or interests in food biotechnology when making regulatory rules

6.POSSIBLE CONCLUSIONS

A full analysis of American environmentalism at the close of the twentieth century may conclude that it is undemocratic because it is occurring during a post-industrial era that is characterized by an administered society. Neither the rational practice of risk nor the liberal theory of politicised environmental risk acknowledges that the current American society is composed of a multitude of social, economic, and political units that control decisionmaking power over the public citizenry (Fisher, 1990). This administered society is not one that responds to the needs, desires, and wishes, ecological or not, of a majority of citizens (Enteman, 1993). In the current age, influence is exercised through elite organizations and, consequently, policymaking responds to whatever the managements of various organizations can gain in their transactions with each other. The future of risk analysis can, however, become more democratically rational because the practice will liberally manage risk. Risk analysis can be more accessible to political processes.

By addressing social values and analytical ethics among public managers, policy analysts, and environmental citizens, I hope that the liberal political theories of social contract can be reasonable arguments to consider in the practice of risk management. Furthermore, such reasonable arguments may offer some insight into the current public debate and protest over agricultural technologies. The merit of investigating rBST as the paradigmatic case study of agricultural risk lies in how it makes clear the complex relationship of public policy and public administration. In conclusion, this complex relationship helps illustrate the problems of the dominant perspective of risk analysis that has incorrectly dismissed significant questions of liberalism in environmental policy.

Democratic decisions on environmental issues are possible only if societal risks have been understood, evaluated, and prioritised in a liberal community. The issue examined here is biotechnology and the use of rational decisionmaking among industry technicians and governmental administrators who dismiss political consent and discourse. At stake are the interests of several players: a biotech company with billions of dollars in investments, small dairy farmers who have adopted controversial techniques in order to be more competitive with agribusinesses, and the majority of American consumers who receive limited information, little choice, and no political power. The lack of public consent or trust about the role of rBST in agriculture is an extraordinary example of the need for a broader debate on risk that communicates citizen trust and appreciates democratic values. Because risk is laden with ethical implications for society, its political analysis must become part of democratic life.

CHAPTER FIVE

A THEORETICAL CASE FOR REASONABLENESS

Solutions for Radical and Mainstream Environmentalists

1.INTRODUCTION

Radical American environmental groups like EARTH FIRST!, Greenpeace, and the Earth Liberation Front (ELF) have claimed responsibility for a list of arsons, defacements, and property damages aimed at the producers, researchers, and proponents of food biotechnology.[20] Since 2000, fervent green activists have executed passive, but illegal, street theatre demonstrations such as those at the annual Biotechnology Industry Organization convention in Boston, staged highly aggressive demonstrations as seen at the World Trade Organization meetings in Seattle, and conducted violent riots at shareholder meetings for the grocery chain Safeway (Martineau, 2001). Whether they are burning crop experiments in research fields at land grant universities in the Midwest or stealing shipments of produce at food markets on the East Coast, their acts of defiance are indicative of the deep ecological concerns some American environmentalists have with genetically modified food (Martineau, 2001; Teitel, Wilson, & Nader, 1999). More important, when Greenpeace activists dumped tons of genetically engineered soy meal onto an American flag outside the biotechnology conference in France or when eco-terrorists destroyed a University of Washington horticultural lab, they were dismissed as ecological anarchists and not environmental freedom fighters by members of the same political movement (Malkin, 2001). That is, most acts by radical ecologists are rejected by mainstream environmental groups because these groups condemn all demonstrations that destroy intellectual capital and improperly characterize the environmental movement as revolutionary.

Despite having a similar political end, many environmental groups have adopted fundamentally dissimilar means to combat the administrative structure that manages environmental risk policy from the perspective of an

[20] Although primary documents, pamphlets, and newsletters, write EARTH FIRST! I will simply type Earth First.

elite and thus incomplete modern liberalism. This division has one significant countervailing effect, mainly that like-minded environmentalists tend to oppose each other rather than the governing political paradigm that creates or allows the public to be at a heightened ecological risk. This chapter describes a hypothetical environmental pressure group that could challenge the dominant public policy formulation process run by economically aggressive bureaucrats who support risk-producing technical industries. This pressure group is created by communication among disenfranchised citizens, activists, and interest groups that all have common environmental concerns. This ambitious movement requires the cooperation of those who choose to stop bulldozers in wilderness areas, who directly lobby political officials, who work to resolve injustice within urban regions, who develop alternative environmental policies, and who try to empower citizens and institutions that traditionally are politically apathetic (Bullard, 1993; Dobson, 1998, 1999a). It is therefore an integrated and inclusive environmental action group.

The purpose of this chapter is to describe a discursive solution to the current division between the two uncompromising American environmental camps. Particularly, in a democratic environment the powerful mainstream forces should not silence dissention by the powerless radicals; instead, responsible public policy should consider both perspectives. This work, using the case study of rBST, will describe the philosophical moves necessary to create the appropriate political atmosphere for policymakers, public administrators, and concerned citizens to ethically debate risk. This inclusive political environmental community would, for instance, seriously attempt to deliberate with the American public about food related risks rather than allow the policy to be created by the elite in government, industry, and science. With regard to the use of rBST, for example, decisions would not be made solely by a science based regulatory system. If contemporary environmental activism concerning the issue of food biotechnology is to be conceived as an inclusive collaboration among its members, it might consider recognizing radical environmental activists as moral disobedients and endorsing risk communication as a vehicle of pluralistic democracy. The intent of this investigation is to propose a new policy framework where numerous discussions of environmental uncertainty and risk management might be organized.

2.POLITICAL AND SOCIAL ISSUES RELATED TO RBST

Although most of the opposition to recombinant bovine growth hormone pales in comparison to the drama created by today's zealous crusaders against "Frankenfoods," established political, business, and academic communities reject both in much the same manner. While some anti-biotech activists have privately staged legal boycotts and illegally dumped harmonized milk in public, the most significant vehicle of rBST opposition over the past twenty years has been the dissemination of information that alleges potential harm to the environment, human health, or animal well-being associated with the use of the new recombinant bovine growth hormone (Steyer, 1997). Countering with industry and government science, however, almost always silences this dissention. For example, during the early stages of rBST approval, the *Journal of Agricultural and Environmental Ethics* published the following:

> Attention to the concerns of the public may be the only means to prevent hysteria over this and future products of biotechnology... Misinformation, from unqualified (and often unidentified) sources pours into the media. This misinformation causes undue hysteria among dairy producers and consumers... It concerns us that the public hysteria leading to the ban on the sale of milk from rBST-treated cows in clinical trials across Canada has effectively brought a halt to much of the industry-funded research in this country... The pursuit of understanding is the essence of science (Burton & McBride, 1989b, 129-130).

This article is not alone, as there are numerous instances of risk managers who attempt to manipulate public criticism and political opposition to rBST by demands for more economic rationality or stronger science. Some of the best examples come from Monsanto employees. Monsanto president Hendrick Verfaillie, for instance, once stated, "We have to convince the consumer that this is good for him," and Monsanto vice president Will Carpenter was quoted as saying, "We have to demonstrate to the public that we're acting in their interest" (Kneen, 1999, 63). The grammar usage of executives should not be over analysed, but it should be noted that these corporate leaders did not say, "We have to *explain* to the consumer" or "We have to *act* in their interest." Although this corporate rationality is the result of agribusinesses favouring the producer of crops over the consumer of food, it appears that biotech regulators are attempting to manage social values of concern with science, economics, and law.

The negative impact of this power dynamic is that opponents might be questioning the science, economics, and law of biotechnology because they are reacting to the undemocratic process of risk policy rather than truly distrusting the product of biotechnology. It appears some environmentalists distrust the science of bureaucratic powers because they feel their concerns are being rationally dismissed under the administrative structure of safety control. For instance, many activists contend that there are unresolved problems with rBST. The most alarming human health problems associated with rBST are:

> (1) Strong role in breast cancer; (2) special risk of colon cancer due to local effects of rBGH on the [gastro intestine] GI tract; (3) May play a role in osteosarcoma, the most common bone tumor in children, usually occurring during the adolescent growth spurt; (4) Implicated in lung cancer; (5) Possesses angiogenic properties—important to tumors, some of which secrete their own growth factors to promote angiogensis; and (6)... acute and chronic effects of IGF-1 be determined in the upper GI tract (InternalReview, 1998, 27-28).

This list is a result of the widely available report by the Health Protection Branch of the Canadian government. It showed distinct immune changes, the formation of cysts on the thyroid, and infiltration of the prostate on 20 to 30 percent of the 90 rats tested.

With this information, the Center for Food Safety (CFS) together with more than two dozen other peripheral environmental organizations suggested further research to United States risk analysis experts at a National Institute of Health (NIH) conference in 1995 and filed a petition in December of 1998 to reverse the 1994 FDA approval of rBST (Teitel et al., 1999). Andrew Kimbrell of CFS stated, "We're going to go to the courts and say—you were lied to. . .it was fraud by the agency and fraud by Monsanto in telling the court that there were no human health effects possible from consuming these products made with rBGH-treated milk. We now know that not to be true"(Teitel et al., 1999, 56). Despite such challenges, the FDA has maintained that milk from rBST-treated cows is not significantly different from the milk of untreated cows and, after reassessment by an independent risk evaluating company, declared the 1998 Canadian study to be inconclusive because original data could not be replicated and was not subject to peer-review.[21]

In summarizing the debate over the use of bovine somatotropin, it is important to understand the main positions of individuals who are opposed to the continued use of rBST in the United States and how those positions are rejected, dismissed, and debunked by mainstream experts and organizations. Critics of the bovine growth hormone today mainly contend that governmental officials and university researchers suppressed scientific studies that showed environmental and human health risks associated with rBST because they were too interested in the economic pursuit of new products and technologies or were influenced by grant monies from biotech companies to publish favourable results. The evidence for these claims seems quite convincing. Furthermore, corroborating these allegations are news stories that identify unreported scientific studies claiming that there are

[21] After re-evaluation by the United States' FDA and an independent assessment by Codex Alimentarius for Canadian authorities and the European Union (EU), it was concluded by the American and Canadian governments that all products from BST-treated cows present no hazard to human health, while the EU has called for more tests. The ban on rBST, however, still exists in Canada and the EU because they contend that it is harmful to the targeted animals. This information can be found at The Institute of Food Science and Technology (IFST) Website.

(1) correlations between increased allergic reactions and rBST milk, (2) certain cancers that recombinant milk causes, (3) connections between bovine somatotropin and some new variant of BSE or mad cow disease, and (4) sections of the American population that will be resistant to certain bacteria because of the increased antibiotic use in cows (Matavalli & Rembert, 1996; Morris, 2000).

Most of this evidence is reported by small environmental groups that are supported by testimonials from governmental employees who contend that they were pressured by superiors to approve the controversial drug on unscientific grounds. Two Health Canada scientists, for instance, were illegally reprimanded by their superiors for publicly expressing their opposition to departmental polices and scientific findings that favoured the biotechnology industry (Kondro, 2000). Another example is from Dr. Richard Burroughs who claims to have been fired by the FDA after ordering toxicology and immunology tests on rBST. Burroughs states, "I was told that I was slowing down the approval process. It used to be that we had a review process at the FDA. Now we have an approval process. I don't think the FDA is doing good honest reviews anymore. They've become an extension of the drug industry"(Teitel et al., 1999, 32). Similarly, critics also allege that certain risks were overlooked because there was corporate influence on researchers doing regulatory risks analyses. This is best referenced by a 1994 General Accounting Office (GAO) report on professional ethics that lists three FDA employees who had a conflict of interest with Monsanto. Additionally, the idea of university and governmental officials being influenced by corporate research grants from Monsanto is damning and is best explained in great detail in books like Ingeborg Boyens' *Unnatural Harvest: How Corporate Science is Secretly Altering our Food* (2000) and Brewster Kneen's *Farmageddon: Food and the Culture of Biotechnology* (1999). Although the particulars are complex, the general theme is that powerful multinational companies exercised improper influence on independent regulatory reviews of rBST in many countries.

Institutions and individuals who support the continued use of rBST in the dairy industry refute most of the aforementioned arguments by stressing the need for rigorous scientific research. Immediately after a news story breaks about alleged dangers of the recombinant hormone in milk—whether it be animal well-being problems, environmental threats, or human health risks—the debate turns into a food safety issue which is a topic that is easily won by biotech proponents who have years of trial evidence supported by almost every established scientific community in the world (Meredith & Robertson, 2000). For instance, on December 15, 1998, the American Broadcast Company (ABC) aired a story entitled "Milk Causes Cancer" which claimed that rBST treated milk caused increased concentrations of the primary insulin-like-growth factor (IGF-1) which is associated with some cancers. On the same day of its release, the American Cancer Society, NIH, and the Food and Agricultural Organization of the UN refuted the story (Monsanto, 1998).

In fact, even countries that have moratoriums on rBST sales within their borders and have banned the import of any harmonized milk or livestock cannot do so on the basis of food safety issues because every major international health authority has agreed that rBST is safe for human consumption. The amount of established science is simply too powerful for any unconventional environmental group to refute. Additionally, defiant researchers who go public with accusations that the regulatory review process has been manipulated are challenged by confident bureaucratic officials who claim that the whistleblowers are either disgruntled employees that were dismissed because of managerial incompetence or are isolated scientists who fail to understand the entire scope of the issue in question. In the first case, they have credible records to support poor work performance and, in the second, they have evidence of professional narrow-mindedness regarding the entire review process. For instance, bureaucratic experts rationally argue that scientists need to realize that technology, based on molecular biology modifications, offers progressive ways to modify existing agricultural processes for the good of all.

They also contend that risk management by its very definition evaluates the costs and benefits not only of scientific outcomes but also social and economic consequences. While it may look like an uncalculated risk to one scientific specialist, it is a decision with a low probability of loss for a bureaucratic risk manager weighing all the pertinent variables. Finally, advocates of corporate financing of biotech research argue that it is a practice that: (1) funds most university programs, (2) is documented and publicly released by technical industries, and (3) endorsed by the Federal government. Consequently, "tell all" books like *Farmageddon* and *Unnatural Harvest* are poorly referenced and attempt to cast too wide of a net of conspiracy to ever be respected as valid research by mainstream individuals and groups. As argued, however, what is important is how the environmental movement is divided between those who have adopted mainstream established public policy strategies and those who employ marginally effective political practices of radicals.

3.A REASONABLE MIDDLE POSITION

Many ecological scholars and activists divide the history of the American environmental movement into three political stages (Dowie, 1995; Gottlieb, 1993).[22] The first phase, often associated with the naturalist John Muir, began with the preservationist impulse of the early twentieth century and coincided with the closing of the frontier and small environmental improvements concerning the wilderness.[23] The second period came in the era of conservationist legislation touched off by the left-wing social movements of the mid-1960s and concluded with the right-wing economic movements of the 1980s. This epoch, symbolized by the environmental interest group leader David Brower, is said to be the apex of American environmentalism. It starts with the historic National Environmental Policy Act of 1969, includes numerous Clean Air and Water amendments, contains the progressive Endangered Species Act of 1973, and ends with the powerful Superfund Legislation. The third wave is the current environmental movement, which can be portrayed by the compromising politician Albert Gore. This era of American environmentalism attempts to find conciliation among governmental bureaucrats, corporate executives, and established environmentalists while dismissing most radical green activists (Dowie, 1995, 175-185).

In the present stage of American environmentalism, with regard to the environmental concerns of food biotechnology, there are two distinctive political factions working to shape environmental policy. The dominant faction talks of "sustainable development" and is accepted by mainstream American corporations, universities, and interest groups. It uses the accepted organizational practices of caucusing, petitioning, and lobbying. The ancillary environmental faction speaks the language of "environmental justice" and uses alternative political methods of mass education, boycott, and protest. In general, radical environmental activists are not sufficiently addressed by the mainstream environmental movement itself, which compromises environmental principles with status-quo institutions of power in order to sustain political development. According to radical activists,

[22] Dowie (1995) gives the best specific description of the three waves of the American environmental movement. It should, however, be noted that introduction to *Forcing the Spring. The Transformation of the American Environmental Movement* by Gottlieb (1993) is a more general and inclusive way to interpret the environmentalism of the past and describe the present contemporary movement. Gottlieb's interpretation situates environmentalism as a core concept of a complex set of social movements that first appeared in response to the urban and industrial changes accelerating with the rapid urbanization, industrialization, and closing of the frontier that launched the Progressive era.

[23] It should be noted that any general periodization of history has some problems of simplification because of its practical intent. For instance, it should be noted that John Muir preservationism was refuted by Gifford Pinchot's conservationism.

mainstream ecological organizations are engaging in conservative agendas and, at best, are attempting subtle changes to environmental regulations governing food biotechnology. Although I think radical ecologists are accurate in their assessment of the mainstream environmental movement, they are ill advised in their political agenda by refusing to work with mainstream environmental organizations.

Similarly, mainstream environmental organizations believe that the radical environmentalism of today stems from the distorted and erroneous political practices of the naive enthusiasts of the late 1960s (Easterbrook, 1995). They think non-mainstream organizations should terminate their tragic ecological sentiment, compose their discontent, and stop their violent and destructive politics of opposition. Mainstream environmentalists warn that fringe eco-movements are "out of the general zone of reality" and urge them to foreswear the practice of doomsaying because it puts the credibility of all environmental concerns at risk (Easterbrook, 1995, 371). Although it is important for mainstream environmentalists to discuss their agenda with corporate and government experts, I contend that it must not alienate the non-mainstream grassroots supporters who are working on alternative political practices within society. That is, the isolated policy issue or passionate gathering of some local fringe group today might be tomorrow's carefully coordinated nationwide campaign for policy change. More importantly, to achieve inclusive environmental reform seemingly powerless activists need to be considered, if not consulted, by dominant institutions with recognized political power.

Many political scientists who study environmental policy have argued that every social movement has and needs both a moderate and an iconoclastic faction (Kraft & Vig, 1994; Lester, 1995; List, 1993).[24] Therefore, it is not appropriate to "reinvent" American environmentalism by institutionalising the left "disobedient types" or democratizing the right "inside the beltway types." Each group exists because it resents the other and represents inherently opposed members of the same movement. To this position, I would simply agree and assert that the two factions of the environmental

[24] Lester (1995), List (1993), and Vig and Kraft (1990) all have general discussions of the mainstream-radical debate of contemporary American environmentalism. It is my project to conduct a specific analysis. There is a very broad spectrum of environmentalism in the United States. The more conservative wing of environmentalism is home to business-sponsored groups such as the Club of Rome and organizations like the Conservation Foundation, which seek an accommodation with industrial society and the cooperation of the captains of capitalism. The environmentalist center is represented by groups such as the Sierra Club and National Resources Defense Council, which distrust industry but accept and utilize conventional channels of political action. More radical groups such as Earth Liberation Front, Greenpeace, and Earth First disdain common politics and pursue direct action. The Sierra Club and Earth First are rightly taken as exemplars of "mainstream" and "radical" environmentalism because the former uses lobbying, litigating, and campaign financing and the latter uses education, demonstration, and civil disobedience.

movement should not be combined. Combining the radical and mainstream camps is not desirable because they perform different, yet necessary, functions. Rather, in order to be more effective at combating the current undemocratic character of public policy development, each group should understand and consider the opposing group's political orientation and value.

It is essential to note that almost every aspect of the ecological crisis is debated among conservative and radical environmentalists. According to mainstream environmentalists, for instance, the human population, industrial production, and energy consumption on earth are escalating but can be regulated to a level of sustainable growth. In contrast, radical groups claim that while forest regions of the earth are shrinking and the desert sectors are expanding, humans continue to emit pollution into the air and water, deplete finite natural resources, and produce indestructible wastes. These contradictory evaluations are possible because not only do the different groups stress different issues, some mainstream groups refute the general science of the alternative groups. While one faction may claim that the protective ozone shield of our earth is dissipating and the temperature of our planet is intensifying, another asserts that global warming is only a theory and greenhouse gas data are questionable. Although I cannot explain all of these schisms, it is important to note that my research—which attempts to justify some of the ecological positions of radical groups with regard to food biotechnology as valid, reasonable, and acceptable environmental politics—is only an example of a larger phenomenon.

It is my organizing thesis that some radical groups exist and flourish because bureaucratic, scientific, and corporate experts have complete control of the public policy process of food biotechnology in the United States. This pessimistic depiction can be made more optimistic and policy professionals can be seen as less homogenous if its bureaucrats are divided "between technocrats committed to the established order and advocates for transformation"(Torgerson, 1997, 345). In this context, dissenting environmental activists can relate with policy professionals in coming up with a better social order. That is, while not commingling, the two will no longer exist in total exclusion of the other.

A fourth wave of American environmentalism can be envisioned—a movement that is both politically accepted and democratically organized—that will promote citizen efficacy, adopt strategies that listen to radical dissidents, and encourage the democratic exercise of pressure groups. There are creative opportunities within the ecology movement for building alliances and connections across community, issue, race, gender, class, and political lines (Bookchin & Foreman, 1991). A discourse among the two divergent camps of environmentalists would encourage ecological activists to unite in building a deeper understanding of the natural world, human community and the environmental crisis common to all (Nash, 1989, xi). The political solutions of united environmentalists—both established interest groups and organized environmental dissenters—who fear the uncertain risks of genetic engineering and its food products would be an example of this new environmental communication or discourse. The first step in constructing this united movement is achieved by having established environmentalists respect the moral dissention of their radical splinter groups.

4.THE POLITICAL THEORY OF CIVIL DISOBEDIENCE

Many scholars today are arguing that the first significant environmental dissenter was Henry David Thoreau (Abbott, 1985). In combining his naturalist literature and moral philosophy, one scholar remarks on the "importance of a life led in a communion with Nature and his uncompromising individualism which insists that it is individuals who must reform themselves if society is to be reformed; not that a reformed society will reform them"(Krutch, 1962, 3). Thoreau is noteworthy because his political and literary commentaries are products of his experiential practice. For instance, his incarceration in January of 1848—argued to be philosophically similar to his naturalist experiences at Walden from July 1845 to September 1847—characterizes and symbolizes civil disobedience in a democratic republic (Abbott, 1985).

Civil disobedients throughout history have argued that laws are not a primary burden, but rather a fundamental good for both the individual and the society. That is, they fully understand that each person is bound to obey the laws enacted by his or her government and they accede that there are political reasons for obeying the law in a society, but they insist that there may be stronger moral reasons for disobeying the same law (Fortas, 1972). In fact, political history in general and American political history specifically has given revered, if not sacrosanct, acceptance for the "men" of conscience who defied their governments (Tolstoy, 1967). Unfortunately, it is difficult, if not impossible, for many activists to be justified as moral dissenters of political laws because their actions are not in correspondence with Thoreau's traditional definition of civil disobedience, an individual act of moral objection that is performed in the public space of politics.

Thoreau's limited paradigm of dissent, however, is still used in the discipline of political science and continues to render invisible some contemporary social movements. It is essential to note that if a protestor is considered a civil disobedient, he or she is instinctively a dignified political actor; if not, the protestor's politics are often perceived as uselessly narrow-minded or idealistic at best, and pointlessly anarchic or radical at worst.

Although civil disobedience is not the only valid form of political protest, it is, however, philosophically accepted as the most ethical. At risk within a biased framework of American civil disobedience are: ignoring moral political activists, calling valuable social reformers something else, and denying honourable protesters the moral justification and political history they deserve. Thoreau's limited definition of civil disobedience needs either to be extended so as to incorporate a more inclusive practice of dissent or rejected for a new perspective that possesses a more comprehensive theory of American civil disobedience.

The traditional requirements for civil disobedience are that the act must be individual, public, nonviolent, and that the protestor must be willing to accept the legal punishment attached to the unlawful act. For instance, a recent American government textbook defines civil disobedience as "breaking the law in a nonviolent fashion and being willing to suffer the consequences, even to the point of going to jail, in order to publicly demonstrate that the law is unjust"(Berman & Murphy, 1999, 510). According to some political theorists, there are some problems with the philosophical clarity of this definition of disobedience in American thought. For example, there has been much debate regarding the condition that the protest act must be nonviolent.[25] Furthermore, some argue that classic figures of civil disobedience, like Martin Luther King Jr., were just aggressive demonstrators and not conscientious objectors. Such arguments have prompted some political theorists to argue that no definition of civil disobedience possesses a level of philosophical clarity that entitles it to be universally acknowledged (Deitz, 1998; Held, 1971, 1972).

Most philosophical work on civil disobedience occurred during the American civil rights and anti-war movements of the late 1960s and early 1970s (Arendt, 1972; Ball, 1973; Bedau, 1969; Gendin, 1972; Goldwin, 1969; Habermas, 1968; Hall, 1971; Held, 1971, 1972; Marcuse, 1972; Zashin, 1972; Zinn, 1972; Zwiebach, 1975). Within this body of literature, numerous definitions of civil disobedience are debated with the somewhat final consensus being that civil disobedience is the middle position between two spectrums of collective political identification (Waltzer, 1970, 11-12). That is, theoretical precision about civil disobedience is problematic because it exists as a practice that affirms or critiques the link between civil culture and political society and, therefore, most attempts to resolve this tension would negate the effective character of this ethical

[25] While Thoreau, for example, is ethically ambiguous about the use of violence both Gandhi and King are morally opposed. Furthermore, contemporary theories of civil disobedience like Rawls, Habermas, and Dworkin characterize disobedience as inherently passive because it is defined as a democratic expression of civil society.

political opposition (Cohen & Arato, 1993). The following literature review highlights the work regarding disobedience and the middle position:

> Jurgen Habermas combines civil disobedience and the question of legitimacy when he says that civil disobedience "must remain suspended between legitimacy and legality." (LT, 106) Ronald Dworkin deposits civil disobedience between "ordinary" crime motivated by "selfishness or anger or cruelty or madness" and civil war (including revolution), also using the question of legitimacy to distinguish civil disobedience from the latter. (MP, 105) Ulrich Rodel, Gunter Frankenberg and Helmut Dubiel claim that civil disobedience is a practice that operates between right and might, between the "morally grounded breaking of norms" and "ideological self-empowerment." Jean Cohen and Andrew Arato say that civil disobedience "moves between the boundaries of insurrection and institutionalized political activity, between civil war and civil society" (Donelan, 1995)[26]

Although philosophically accurate, the "in-between" or self-contradictory explanation of civil disobedience presents one significant political problem. Because the theoretical definition of civil disobedience successfully lies between established universals, the only means for justifying an act as legitimate dissent relies upon an outside perspective. That outside perspective, due to its subjective power, must be an inclusive arbitrator in order to consider the voluminously different acts of political dissent.

While it is important to distinguish a revolution from a protest and an inherently criminal act from an explicitly political one, it is more important that such a distinction is made from an unbiased perspective (Rucker, 1966). As discussed, most attempts at political distinction lead to a list of assumptions that the "disobedient act be public, deliberate, and illegal," which traditionally results in definitions that use classic cases of dissenters like Thoreau, Socrates, Jesus, Gandhi, and King as the moral guide to ethically discuss other acts of protest (Power, 1970, 37). For instance, regardless of the scholarly work that negates a simple theoretical definition of civil disobedience, Henry David Thoreau's essay and event continue to mistakenly embody the concept in American thought. The intent of this section is not to propose a new definitive figure of civil disobedience, but to expand the perception of civil disobedience as an effective form of political opposition for many diverse individuals. Although I am not attempting to create a universal definition, I am working toward a description that is not as prejudiced as the traditional notion and is more helpful than abstract philosophical deliberations.

[26] Please see Donelan (1995). The referenced quote is taken from pages three and four in which he references Jurgen Habermas, 1987. "Civil Disobedience: Litmus Test for the Democratic Constitutional State," *Berkley Journal of Sociology*. 85: 95-116; Ronald Dworkin, 1985. *A Matter of Principle*. Cambridge: Harvard University Press; Ulrich Rodel, Gunter Frakenberg and Helmut Dubiel, 1989. *Die demodrtisch Frage*. Frankfurt: Suhrkamp Verlug, 25; and Jean L. Cohen and Andrew Arato, 1992. *Civil Society and Political Theory*. Cambridge: MIT Press.

Despite the many definitions and cases of civil disobedience, most include Thoreau's requirements that the act be protested individually and in public. But a myriad of experiences by women in the democratic polity has prompted some scholars to assert that the only common characteristics of civil disobedience are that the act be a violation of some law, and that the protest be performed on the grounds of social conscience rather than for political gain. This perspective has been defended in many places, and is, I think, persuasive (Sparks, 1997). The male-focused mainstream paradigm that insists that an act of civil disobedience must be protested individually and in public has to be discarded despite the fact it is found in some of the most sophisticated philosophical work on civil disobedience. Furthermore, it is necessary to refute the personification of Thoreau as the defining civil disobedient because subsequent environmental dissidents often protest in a private style rather than a public one and are better understood through communal rather than contractual political theory.

For instance, there are cases of a single woman or group of women who attempt to change public officials' opinions on ecological safety or policy positions on technological risks to their family or community by climbing aboard nuclear submarines, boycotting certain products with food strikes, writing fraudulent letters to decisionmakers, or organizing groups for direct action (Mollin, 2000). They provide examples of traditionally personal practices that have political consequences (Shiva, 1994). Perhaps this is best seen in the following narrative. Lois Gibbs, an ordinary suburban housewife, emerged in 1978 as one of the most unlikely environmental activists when she discovered that her Niagara Falls neighbourhood, Love Canal, was built on an industrial landfill left behind by the Hooker Chemical Company. After hearing neighbours' complaints about their children's respiratory problems, homes with mysterious black silt in their basements, pets losing large amounts of hair, and adults with unusual skin irritations, Gibbs recorded all the abnormalities and organized the Love Canal Homeowners Association to investigate possible causes (Seager, 1993, 264). Despite having Hooker executives and governmental officials discredit Gibbs's claims as "useless housewife data" and "hysterical housewife" reasoning, she continued to pursue political action by almost any means necessary (Seager, 1993, 265).

After continued argumentation between Gibbs and local stakeholders, university scientists visited the community of Love Canal and discovered high rates of birth defects, miscarriages, epilepsy, liver abnormality, and headaches (Dowie, 1995, 128). Following that, the New York State Health Commissioner finally announced that the landfill in Love Canal "constitutes a public nuisance and an extremely serious threat and danger to the health, safety and welfare of residents"(Dowie, 1995, 128). Despite the public declaration, state and federal governmental actors failed to take any actions to aid Gibbs and the citizens of Love Canal. As a consequence of governmental inaction, in the summer of 1980 when two federal inspectors from the Environmental Protection Agency arrived to inspect Love Canal, Gibbs and other residents held them hostage. Two days later, President Carter arrived in Love Canal, New York, declared the area an uninhabitable toxic zone, and agreed to purchase homes and relocate residents affected by the national disaster. Today Gibbs is heading a national organization entitled Citizens' Clearinghouse for Hazardous Wastes, which provides counsel and support to thousands of small grassroots environmental activist groups.

Gibbs's unyielding persistence, illegal activism, and political initiative have shaped the nature of grassroots environmental politics in America, especially for women (Seager, 1993, 265). In the general language of trust rather than the specific language of risk analysis, Gibbs once questioned the probabilistic science of governmental regulators when she rhetorically asked in a speech: "Would you let me shoot into a crowd of 100,000 people and kill one of them? No? Well, how come Dow Chemical can do it to us" (Gibbs, 1983)?[27] In fact, Gibbs's movement stands out as one of the most effective women's community-based environmental organizations in industrial history. More important, however, Gibbs's practice of resistance in the Love Canal battle embodies one of the most important theoretical characteristics of a feminist view of civil disobedience. Although some modern acts of disobedience are not understood in the conventional public space and traditional style of protest, they are still very political. Whereas Gibbs's meetings were held in her living room and a lot of her communications were transcribed in her diary, her motivations were not private self-interest but public welfare.

In order to better understand the second integral characteristic of a more inclusive understanding of American civil disobedience, it is helpful to examine other environmental protests in America. Among biotechnology opponents, Jeremy Rifkin is the modern day Henry David Thoreau. As early as 1983, Rifkin and his Foundation on Economic Trends (FET) disputed the advantages and proclaimed the destructiveness of biotech corporations, most especially Monsanto (Comstock, 1990). As a radical environmentalist, Rifkin chose to educate the public about the worst-case scenarios of what his followers called "Frankenstein-Foods." The results of his efforts are seen in the many consumer, animal, agriculture, and environmental activists who became interested in the emerging technology of genetically altered food and its first commercial product, rBST. In fact, Rifkin claims to have been the

[27] This quote is attributed to Louis Gibbs as a speaking quote but is not directly referenced in Gibbs (1983).

single individual to "make BGH an issue" for environmental opposition (Charles, 2001, 94). For instance, the Humane Society of the United States became interested in the animal welfare of the treated cows; the Agricultural Resources Project under Jack Doyle questioned the research sponsored by genetic engineering firms; and the Wisconsin Family Farm Defense Fund (WFFDF), the National Save the Family Farm Coalition (NSFFC), American Agricultural Movement (AAM), National Farmers Organization (NFO), and National Farmers Union (NFU) collectively argued that rBST would create fewer and larger farms in the United States (Browne & Hamm, 1990).

During a series of regional public hearings on rBST, a panel of FDA officials listened to radical environmentalists, like Michael Fox, who contested the new technology. His full testimony reads:

> I wish to go on the record as being opposed to FDA approval of genetically engineered recombinant bovine growth hormone (rBGH) until there is a valid assessment and resolution to the mastitis (infection of the cow's udder) issue. I have reviewed published reports on the effects of rBGH on commercial dairy cows. These diseases, especially mastitis, will be predictably exacerbated by rBGH treatment. The drug, rBGH, essentially hyperstimulates the dairy cow to produce milk, which can lower the animal's disease resistance and contribute to increased incidence of lameness and mastitis. Profound physiological changes can also lead to metabolic disorders and reproductive problems, most notably infertility. We think that the incorporation of rBGH into dairy cow husbandry is likely to create an opportunistic niche for viral and mycoplasmas disease. This could become a costly endemic problem in the nation's dairy herd, especially in view of the potential immunosuprevive effects of rBGH. Alternative husbandry practices, such as rotational grazing, will increase productivity without jeopardizing dairy cow health and welfare. If FDA approves rBGH, we will consider supporting a nationwide boycott of all dairy products until FDA mandates labeling of products derived from cows' not injected with this hormone. There are also adverse socio-economic consequences to small-and midsize dairy farms. We cannot overlook public health concerns (especially over the unresolved question of insulin-like growth factor in rBGH milk). There are consumer right-to-know issues arising from the specter of rBGH being injected into the nation's dairy herd. We should not overlook the most obvious economic inefficiency of using rBGH to increase milk production. When farmers use rBGH they have to feed their cows more costly high-energy concentrate feeds. This wastes the natural ability of the dairy cow to efficiently convert low nutrient plant material into milk protein for human consumption. The incorporation of any genetically engineered product like rBGH into the food production system is a backward step indeed. Finally, the public equates milk with purity. Public support will erode if the dairy industry adopts the use of rBGH. The public is well aware that antibiotic milk-residues will become more of a problem because of an increase in mastitis in the national herd paralleling increasing use of rBGH. Clearly, rBGH is a product that is beyond the FDA's scope of objective, science-based evaluation. Its approval rests upon a regulatory framework that in theory purports to protect consumer health and safety, but in practice would be so costly to effectively enforce. Costly enforcement at public expense would negate any purported saving to the public of a cheap and plentiful supply of milk from rBGH-treated cows (Fox, 1997, 96-98).

In the testimony of Dr. Fox, the populist aversion to the corporate domination of food biotechnology, the unreasonable demand for risk-free or natural food or the call for direct democracy decisionmaking is missing. This is noticeably different than contemporary anti-biotech companies who refer to rBST as "Crack for Cows" and criticize Starbucks for not using organic milk at their coffee shops. Instead what is noticed in Fox's thought is a logical argument for employing ethical considerations in developing food biotechnology companies.

Michael Fox, while still a radical, articulates his positions in a very just fashion. They are: (1) a "Genetically Modified" label for rBST milk, (2) an open and transparent regulatory agency for food biotechnology, and (3) a public communication about risk that is not scientifically based. Fox not only articulates areas of agreement between social activists questioning biotech food and mainstream environmental groups lobbying governmental agencies about the regulation of genetically modified organisms, he seems to embody a new type of disobedience. With regard to moral protest, there are three important things to be discerned from Dr. Fox's testimony. First of all, like many mainstream environmentalists interested in food biotechnology, Fox prefers to use the term recombinant bovine growth hormone or rBGH rather than the scientific one of recombinant bovine somatotropin or rBST. Secondly, his argument does not use statistical rationality, but instead reasonable theorizing. That is, the ethical problems associated with animal welfare risks, unresolved human health issues, and significant oversupplies of milk are all problems that can by discussed without the exclusive use of scientific evidence that silences ordinary citizens. Finally, the politics endorsed are untraditional because boycotting a product on the grounds of ethical values is more akin to the radical illegal practices of civil disobedience than to the traditional legal maneuverings of environmental interest groups. This is significant because Fox's opposition is an organized communal dissent not an individual contractual one.

There are generally three divergent voices regarding civil disobedience: one speaking for absolute abolishment of civil disobedience, a second advocating limited acceptance, and a third justifying civil disobedience in a very conservative manner. In the first case, one assumes that society and government are always ill served by acts of disobedience. From this perspective, any individual who publicly disobeys a political law, immorally risks the chaos of civil anarchy for the entire society (Ball, 1973). The second perspective maintains that legitimate grounds for civil disobedience are contingent upon the system of government from which the unjust law derives. In this case, disobeying the law is accepted under certain political environments; for instance, it is advocated in autocratic regimes, but never accepted in a democracy. The logic behind this argument is that no individual should disobey a law in a community where the opportunity exists to change the law through an election or any other political participation (Scott, 1990).

The third voice argues that civil disobedience is sometimes justifiable and occasionally beneficial under any political environment. However, it does not necessarily follow that one is freed from the obligation to obey a law simply because it is a bad law. The rejection of an unjust law is only justified if the obligation it imposes is immoral and if individual dissention can have public consequences. In a very conservative tone, this voice asserts that the consideration of whether to obey is a very serious obligation. One may illegally protest a law if legal opposition was unsuccessful and the person is subject to continued, deliberate, and frequent violations of civil liberties (Rawls, 1971a). In this view, within a democratic state, resistance is primarily an individual conscious problem that is vindicated in very rare and public occasions. This third and only liberal justification of civil disobedience in political science ignores many of the protests of local environmentalists in a democratic republic.

Its claim is that, after exhausting all legal alternatives, the most common characteristic of dissent is that the action of civil disobedience be by an individual in a public space (Hall, 1971). In fact, this third understanding of civil disobedience constrains radical environmentalists in the same manner—if not more severely—than the previous two perspectives of civil disobedience. By refusing to consider the private space and style of action that encompasses many environmental protest theories and numerous feminist practices of civil disobedience, this justification does not adequately include the whole range of possible actions of civil disobedience. Since women's access to the public has ranged from limited at best, to forbidden at worst, it is impossible to recognize most of their acts as civil disobedience in the traditional sense (Sparks, 1997). Furthermore, because environmental protesters are often new political actors and opposed by the dominant social, political, and economic interests of society their practices of opposition are untraditional either by design or function.

Many feminist thinkers draw from women's experiences in the familiar sphere of intimate and domestic connections to argue that the formal "public" politics of the democratic community must consider related "private" notions of power (Morgan, 1996). In theoretically asserting that women's actions can be found less in the formal performances of politics than in the informal realm of power, one must consider the very limited scope of the notion of civil disobedience and its related ideas of justice, power, and order. As a consequence, many feminist political theorists argue that liberal democracies need feminist communities to help develop citizens, solve collective problems, and legitimatise decisionmaking processes (Mansbridge, 1993). Such a reality has provoked feminist theorists to wonder what a women-focused account of protest would reveal (West & Blumberg, 1990). Similar to this end, in the next section I will attempt to reconstruct civil protest from the communal perspective that, unlike the contractual, would include women's social movements and modern environmental ones.

The mainstream American environmental movement needs to think of radical dissenters as civil disobedients promoting the liberation of exploited and oppressed members of the American ecological community. If radical activists are viewed as civil disobedients rather than eco-terrorists, mainstream environmentalists will be more likely to cooperate with them and the two groups can jointly facilitate a more democratic environmentalism. Mainstream environmentalists should note that in adopting a subversive and counterculture stance, radical ecologists have focused on the quintessentially liberal foundation for protecting nature: natural-rights philosophy, the old American ideal of liberty applied to nature (Nash, 1989, 12). Although other kinds of environmentalism value goods that cannot be reached by social contract politics, this type of environmentalism is about communal trust and responsibility. Seen as the embodiment of American politics rather than its abomination, philosophical acceptance is crucial for justifying the politics of radical environmentalists.

For instance, the radical interest group Earth First should be seen as "freedom fighters" battling environmental racism rather than as "terrorists." For this to happen, Earth First's strategy of monkeywrenching—interfering with industrial and commercial processes that damage the environment—must be seen as something that reinforces the democratic process (R. Young, 1995). Monkeywrenchers seek to slow traditional technological advancement by getting political attention focused on the need for environmental caution within the modern liberal concern for progress. Critics claim this to be a form of needless terrorism that privileges trees over landowners. But, supporters equate it with civil disobedience. Monkeywrenchers take risks because their actions are grounded in moral convictions that have the practical consequence of mobilizing public opinion (Lee, 1995). The power of civil disobedience in the American environmental movement should be considered.

As noted earlier, acts of protest can be justified as instances of civil disobedience if done in good conscience for a collective benefit. As long as the act is a violation of some law and is performed on the grounds of conscience rather than merely for personal gain it deserves the respect and attention of those who engage in mainstream political practices.[28] That is,

[28] Please note that this definition is purposefully lacking any discussion of non-violence. This omission is because I would like to include protestors who in demonstrating the unjust nature of some environmental regulations, have participated in breaking the law in a violent fashion. I would, however, observe that all acts of environmental terrorism are targeted toward things and not people. This distinction is noteworthy because it is used by federal law enforcement agencies for categorizing terrorist organizations.

some untraditional acts of environmental protest are acceptable in the modern age when radical or individual voices fall on organized deaf ears. Protestors of environmental racism are a good example of organized communities of dissent worthy of ethical acceptance. Environmental justice—resolution of ecological problems requires fundamental social change because the benefits of environmental exploitation accrue to the wealthy class while the poor and the marginal bear the costs—is important to this discussion because it explains how grassroots groups can defy the dominant American movement that silences dissenting voices (Bullard, 1993). In many instances, grassroots leaders emerge from groups of concerned citizens who feel threatened by some type of governmental policy. These groups and their leaders embody a new political power and social order in today's administered society. In fact, some of these grassroots leaders have forced their issues onto the nation's environmental agenda.

Although she is not an environmentalist author, the political philosopher Hannah Arendt explains the lost purpose of politics in the malaise of the post-industrialist society:

> "Wherever you go, you will be a polis:" these famous words became not merely the watchword of Greek colonization, they expressed the conviction that action and speech create a space between the participants which can find its proper location almost any time and anywhere. It is the space of appearance in the widest sense of the word, namely, the space where I appear to others as others appear to me, where men exist not merely like other living or inanimate things but make their appearance explicitly. . .Although all men are capable of deed and word, most of them—like the slave, the foreigner, and the barbarian in antiquity, like the laborer or craftsman prior to the modern age, the jobholder or businessman in our world—do not live it (Arendt, 1958, 198-199).

Here, Arendt analyses the nature of human society as it has developed in the modern age. For Arendt, generally, the linkage of true politics and personal reality is the salvation of the crisis of the human condition in today's egocentric republic. Specifically, for Arendt, politics is polis in Greek and civitas in Latin and therefore is the craft of civilization together with the necessary events of free community. Arendt offers a theoretical bridge from the administered individuals of private political actions to the communitarian liberals who organize their antagonism in order to make a political statement of conscience. The ideal democratic political arrangement for the contemporary age needs structure, but accepts an open society because its creative citizens would be characterized by communal tendencies. This type of political thought could improve environmental policymaking in America.

Developing political theories of the public policy process is an important enterprise within political science given the present undemocratic character of environmental policymaking in America. This is best seen in the development of environmental concerns regarding food biotechnology in the

United States over the past twenty plus years. As early as 1978, agricultural and economic opponents of the first biotech product for human consumption, rBST, began to emerge (Comstock, 1990). Their objections to the contested technology have evolved from social concerns to include economic impacts and environmental issues. While environmental attitudes toward rBST were originally neutral they quickly became cautious as Monsanto created the Industrial Biotechnology Association (IBA) to lobby the FDA, EPA and OCED (Organization for Economic Cooperation and Development) on the impact of biotechnological innovations generally and the genetically created bovine growth hormone specifically (Browne & Hamm, 1990). As already seen, opposition to the development of rBST waxed and waned during its decade long regulatory review process, and today it is not much more than a radical environmental concern. Mainstream environmental groups should have learned the importance of participatory policymaking in risk management from the first case of food biotechnology. For instance, the lesson of pluralism from the dairy debate over rBST has been a significant issue in mobilizing concern about Bt crops.

5.THE PRACTICE OF PLURALIST DEMOCRACY

Shortly after the approval of rBST in 1994, discourse about the risks of food biotechnology was a viable topic only among the radical fringe of the American environmental movement. But with the chaos created by European anti-biotech protests came mainstream American environmental acceptance of many of the position statements of obscure organic farm organizations, progressive economic interest groups, and radical environmental disobedients. Today, for instance, seven organizations—including mainstream environmental veterans like the Sierra Club, Friends of the Earth, and the Public Interest Research Groups—are launching the Genetically Engineered Food Alert. It is a million dollar, multi-year, multi-participatory, effort to pressure Congress and the FDA to force the pre-market safety testing and labelling of all genetically engineered organisms found in milk and all food stuffs (Roosevelt, 2000).

It is only through viewing radical environmentalists as civil disobedients while maintaining an openness toward participatory models of public policy development that the traditional notion of pluralist group politics can be used for describing the current undemocratic situation of food biotechnology regulation and, more importantly, for prescribing a more liberal one. Major American newspapers, like the New York Times, have been running serial articles about the growing concerns regarding consumer acceptance of biotech crops. An interesting theme in these articles is that many of the consumer and environmental activists are not concerned with the food safety science of genetically modified food and frequently do not oppose their research and sales in the formal, traditional, or conventional manner. Unlike the rBST dissenters who attempted to refute industry and governmental experts with opposing data, Frankenfood protestors resort to alternative political measures. Although scientific engineering may be essential to discussing the use of biotech crops in modern farming, a coherent set of ethical principles for political debate cannot be developed solely on scientific comprehension (Reiss & Straughan, 1996). And, the government's continual attempt to do so will be countered in dramatic fashion.

What appears to be a scientifically sensible conclusion regarding the use of genetically engineered plants and animals in the U.S. is beginning to be seriously scrutinized because consumer unease has coalesced into consumer activism (Bykerk & Maney, 1991). From the Boston Tea Party to the Stamp Act Boycotts to the Civil Rights Wage Equality Walkouts, consumer activism has typically developed in settings when American citizens have restricted access to power and formal governmental channels (deGrazia, 1996). When protestors have little recourse through conventional political arenas, they use extra-governmental or market channels to influence governmental decisions. Consumer activism is not the only means to protest, but it is an important one for many radical environmentalists. An important element to understand in most consumer and political protests against food biotechnology is that, despite the fact that the protestors operate outside of conventional scientific rationality and traditional political practices, their message needs to be understood by both conventional stakeholders and traditional policymakers. Therefore, in order to be politically relevant, radicals must be heard in a familiar and recognized space. That is, "greater democratic control of environmental risks that the public faces would ensure" the societal goals are discussed in a pluralistic manner rather than an elite one (Thiele, 2000, 559). Perhaps this is best illustrated by environmentalists who are wary of technological and economic growth because society seems to treat them as inherent goals rather than small means to a greater end.

It is essential to note that the modern liberalism desired in contemporary risk analysis is not a radically new ideology, only a more reasonable environmental understanding of the old modern liberalism. Adapting participatory models of policy planning that use wider strategies and a democratic ethos, as argued, may alleviate the loss of the American ideal of democracy in the post-industrial age. Furthermore, it has been illustrated by environmental justice scholars that despite their subversive activities, radical environmentalists' operate from the quintessentially American foundation of natural rights. Understood not so much as a revolt against traditional politics but as an extension and new application of them, radicals will be accepted by the mainstream as justified civil disobedients. The way to achieve this is to have environmental interest groups integrate their pressure politics. Thus, if mainstream politics considered rather than dismissed radical politics, policy sciences might realize that their discipline could be the perfect blend of democratic theory and practice.

American public policymaking aspires toward democratic theory, but it must function between the parameters established by several accepted political, social, and economic principles. The cornerstones of contemporary democratic theory are explained by Robert A. Dahl, who suggests that an ideal democratic process would satisfy definite criteria: "equality in voting," "effective participation," and "enlightened understanding," which rests on a system with free speech and free press that creates a marketplace of ideas where civic understanding is not monopolized or distorted by a single dominant paradigm (Dahl, 1983, 6). Similarly, the last two distinguishing factors of a democracy are "final control over the agenda," meaning citizens have a collective control in setting the policy agenda to be debated, and "inclusion" of all under the government's rights (Dahl, 1983, 7). Although Dahl is a traditional democratic theorist, his depiction of an ideal democracy can be used to develop a progressive ecological politics that enables citizens to act together effectively and pursue shared objectives both in a sustainable and just environmental fashion.

Only by following these principles can a system qualify as being "democratic". Almost all Americans believe in normative democratic theory. The question, though, is how well modern American democracy empirically lives up to these high ideals in practice. Direct control by the citizens is replaced by an indirect relationship between a few leaders and many followers; American political science refers to this as substantive representation. More importantly, the closer the substantive correspondence is between elite representatives and their mass citizenry, the stronger the democratic republic (Lineberry & Edwards, 1989). There are, nonetheless, many reasons that representatives, both elected officials and career bureaucrats, fail to mirror majority preferences and thus undermine a strong democracy. As such, the system of representatives and their democratic power is often altered by small groups who are competing for specific interests. Being suspect of the dangers of interest groups to democracy is warranted because not all groups are created equal in power, membership or resources.

With regard to American environmental policy, one of the main reasons for weak democratic representation, poor citizen participation, and a lot of bureaucratic cynicism is the unequal distribution of power. More important than the cause is the solution. To enjoy effective representation, make democracy work, and redistribute power, most environmental scholars endorse the public policy effect of pluralism or competing interests groups. Pluralist theory rests its case on multiple centres of power in the American political system. Most political scientists argue that the extensive organization of competing groups is evidence that political power can be widely divided. Like parties and elections, interest groups are a linkage institution between governmental experts and the citizen at large. Interest groups are an established form of political practice in the United States that pressure government to support some public policies or reject others. Although the American public have long suspected that interest groups exert corrupting influences on our representative political system, policy scholars argue these accusations eclipse the potential of competing groups.

Mainstream or public interest groups claim to be different from radical interest groups, representing, they say, an important part of the public interest. Radical or single-issue groups differ from traditional interest groups in that they are narrowly focused on one issue on which they refuse to compromise.[29] Interest groups play a role in every theory of American government, and democratic pluralists argue they are the most important way people can have their policy preferences represented in government. According to pluralists, you win rarely and you lose frequently, but no single group should lose all the time. In effect, a group, whether mainstream or radical, is never shut out completely from the American political system. The group theory of politics offer the following essential arguments: groups provide a key linkage between people and government; groups compete; and a rough approximation of the public interest emerges out of the multiple group struggle for governmental power and policy influence (Lineberry & Edwards, 1989).

The phrase "public interest" is so frequently made reference to that one might conclude that every political interest group worked to pursue it.[30] The phrase suggests that there are general interests superior to the private interests of individuals, concerns we all have in common. Public interest lobbies have been defined as organizations that "seek a collective good, the achievement of which will not selectively and materially benefit the membership or activists of the organization"(Berry, 1977, 7). In practice,

[29] It is important but not essential to note that some interest group scholars would not consider radical groups single-issue groups. I would assert, however, this conclusion is not valid because it is operating under the notion of rational modern liberalism not reasonable modern liberalism.

[30] The terms public and special interest are of course more rhetorical than practical. That is, some mainstream groups like manufacturers associations have very non-public interests and some special interest groups like consumer advocacy coalitions have very public concerns. These educational terms are used to demonstrate the inclusive nature of environmental interest groups. Most contrasts, like the many used in this chapter, are false dichotomies that attempt to make each other's politics exclusive.

however, it appears that "public interests" are those things advocated by mainstream organizations while "special interests" are those advocated by radical groups. While difficult to measure, there are approximately 3,000 interest groups concerned with the environment and they are ubiquitous within all areas of environmental activity and issues.[31] Although most groups claim to represent some version of the public interest, in reality most green pressure groups are considered single-issue groups.

Single-issue groups also use the policy tactic of pressure politics, but differ significantly from traditional interest groups in both the scope and depth of their interests. Single-issue or radical groups have three general features: their interest is narrow; they tend to dislike compromise; and they draw membership from people new to politics (Lineberry & Edwards, 1989). Radical environmental groups are concerned that the timid campaigning of mainstream groups is doing too little too late to save the global ecosystem. Radicals suggest direct action as a strategy for opposing energy production, industrial expansion, and wilderness destruction. Their activities have drawn criticism from both inside and outside the green movement, and they are variously accused of valuing animals and trees above human beings, of endangering human life, and of giving the rest of the environmental movement an anarchical name.

Radical environmentalists want to change the world but they, like Henry David Thoreau, maintain that societal change is achieved through individuals not institutions. They want increasing numbers of people to experience nature, industry, farming, or economics in terms of a new ecological sensibility. According to radicals, an environmental evolution must occur, similar to the agricultural and the industrial paradigm shifts of world history that revolutionized human existence, in order to render the Earth sustainable for future generations. They argue that such an environmental evolution will lead inevitably to economic, societal, and political structures radically different from those of the present. Like mainstream public interest groups, radical special interest groups use pressure tactics in order to influence public policy making. Unlike mainstream organizations, however, radical groups construct strategies that reach beyond rational politics. The following document explains:

> Finally, the actions of monkeywrenching invariably enhance the status and bargaining position of more 'reasonable' opponents. Industry considers mainline environmentalists to be radical until they get a taste of real radical activism. Suddenly the soft sell of the Sierra Club and other white-shirt-and-tie eco-bureaucrats becomes much more attractive and worthy of serious negotiation. These moderate environmentalists must condemn

[31] Although the exact number is debated, many authors of introductory environmental policy textbooks report a number similar to this one.

> monkeywrenching so as to preserve their own image, but they should take full advantage of the credence it lends to their approach (Abbey, 1990, 76).

Radicals believe that achieving a sustainable environment depends on opposing the engine of dominant political activities of American culture, so change must go beyond the traditional politics of rational arguments (Dryzek, 1997).

Although untraditional in their pressure politics, radicals are quite helpful to the conventional green political process. According to radicals, if one wants to convince the public of the fragility of nature or the risks of technological innovation, then that has to be done by reaching the listener through reason and not rationality. Radicals reject the conservative indirect action of traditional politics and embrace direct action. Nevertheless, the point of their action is the same as that of more traditional groups: both wish to exert pressures that shape the development of public policy.

Radical organizations, a noteworthy category of special interest groups, are perhaps known less for their policy objectives than for the pressure politics of their members. "These exploits include lying down in front of bulldozers, gatecrashing the anniversary celebrations of Lake Powell reservoir on the Colorado river, occupying the tops of trees in old growth forests scheduled for clear cutting, putting a `crack' on the face of the Glen Canyon Dam, and so forth"(Dryzek, 1997, 168). Like all civil disobedients, environmental protesters are not afraid of legal consequences. For instance, recently a woman perched herself atop a California redwood tree to protest old growth logging. In 1994, over twenty Earth First members were arrested for creating a two hour blockade on a road leading to a Tennessee nuclear plant (Cockburn, 1996).

Earth First is known for monkeywrenching or eco-terrorism. Earth First invented eco-terrorism, a practice used today by many radical environmental groups including Greenpeace and the ELF. Eco-terrorism is described as direct political engagement, like "throwing a wrench into the gears" of an industry that is polluting or harming Nature, in order to immediately stop the destructive behaviour and/or receive media attention for that environmental cause (Eagan, 1996). As mentioned, monkeywrenching plays a significant role in the rhetoric of Earth First supporters. Dave Foreman's field guide, *Ecodefense*, includes the following examples of eco-terrorism: pouring emery powder into the crank-case of earth-moving machines, hammering spikes into trees, pouring syrup into fuel tanks, pulling up survey stakes, and destroying logging roads (Foreman & Haywook, 1989). It is essential, however, to note that the number of actual cases of environmental terrorism by radical green groups does not match up to their dramatic claims. That is, monkeywrenching is used more as a threat than as a tactic and the best accounts remain those of the deep ecological nature writers and theorists (Dryzek, 1997).

The rhetoric, fiction, and philosophy of radical environmental groups are, however, very important. As discussed, "Earth First activists still wish to raise the level of confrontation between mainstream environmental. . .countermovement, corporations, and the State. Earth First members advocate no compromise in the defense of Mother Earth and engage in extra-institutional activism, including nonviolent civil disobedience, strategic property destruction and other forms of direct action"(Ingalsbee, 1996, 241). Perhaps, the most effective force of the radical groups today and throughout history is their romantic notion of environmentalism. More than anything else, radical activists are green romantics who are apathetic about social structures and institutions of industrial society. As a consequence, radicals have little to say about governments, markets, and policies. Green romantics can justify this lack of interest on the grounds that social structures, institutions, and policies have no life of their own, but are ultimately reducible to the underlying sensibilities of the members of society (Dryzek, 1997). For instance, environmental justice activists reject science, embrace empathy, and cultivate radically different ways of experiencing oneself, urban life, and nature. For these radicals, politics is not about devising strategies to achieve tangible goals by asserting objective truths, but it is an arena in which different subjective experiences can be expressed. Furthermore, romantic green sensibilities exist beyond its radical members (Pinkerton, 1997). For instance, the philosophy of green romanticism is also in the political theory of mainstream groups.

Green romanticism may provide the real impetus for the environmental activism of some group members and leaders, even as they may be forced to suppress these sensibilities when they enter the arena of conventional politics. In fact, many green romantic individuals operate in both the conservative and the radical green pressure groups. The recently deceased David Roth Brower was, perhaps, the most visible, militant, and influential figure in environmentalism in the United States since the 1950s. Green romanticism is very evident in his speeches as the leader of the Sierra Club and in his foundational writings for the League of Conservation Voters, Friends of the Earth, and the Earth Island Institute. Because green romanticism is used by mainstream and radical environmentalists it might have the potential for uniting the current divide in the third wave of environmentalism and offering inclusive policies for the fourth wave of green politics.

6.CONCLUDING REMARKS

The divergent factions of environmentalism discussed in this section—mainstream interest groups and radical social organizations—can cooperate with regard to the issue of environmental injustice with the praxis of policy-interest pressure. Perhaps this context of agreement can offer suggestions for improved policymaking in a fourth wave of American environmentalism.[32] In fact, some environmental roundtable conventions have recently been created in order to bring together multiple green-oriented groups that have failed to cooperate in the past (Baxter, 1997b). Efforts to combat environmental injustice provide an example of how mainstream and radical interest groups can work together to achieve desired changes.

Environmental injustice refers to the fact that poor communities are disproportionately harmed by pollution and that corporations and government, intentionally or unintentionally, build their worst environmentally hazardous facilities in and around these low-income neighbourhoods (Rosen, 1994). It also refers to the well-documented fact that local governments have excluded minority communities from environmental planning and that toxic sites have damaged many traditional minority communities. Mainstream and radical groups have occasionally cooperated and provided minority organizations fighting environmental injustice with invaluable expert testimony, research, technical assistance, and legal advice. Recently, radical groups have demanded support from their mainstream allies. One strike came in the form of a publicized letter in which a small grassroots organization bluntly demanded that mainstream environmentalists "look at how the cultural bias of an all-white green movement shaped many traditional environmental policies and goals" (Baxter, 1997a, 101).

In a very real sense, these environmentalists come from and are concerned about very different worlds. At best, mainstream groups focus on air, water, and land—the health of a planet that cannot protect itself. In contrast, the movement for environmental justice, represented by radicals, is not concerned with overabundance, but with environmental, social, and economic inequalities that unequally disadvantage traditionally non-political actors. Some mainstream environmentalists charge that minority grassroots organizations are not sufficiently concerned with the environment, and only seek a fairer distribution of environmental dangers. Furthermore, when poor

[32] Please note that my description of a possible fourth wave of environmentalism is not an activist manifesto for a prescribed grassroots environmental movement. I have not suggested that if grassroots citizen participation was passionate enough, it could achieve sustainable development, alleviate poverty and hunger, eradicate disease and achieve state self-sufficiency. Rather, I have argued that the theory of sustainability could be greatly enhanced with the practice of environmental justice. In fact, many scholars believe that the key to future success lies in integrating all aspects of the environmental movement.

communities accept monetary reparations from polluting industries, mainstream environmentalists charge them with being "bought off." These allegations, however, miss the potential of a cooperative environmental movement. Radical and mainstream environmental groups that work together are informal institutions that greatly influence the politics and government in America.

The combined mainstream and radical battle against the issue of environmental injustice is important because it embodies diverse grassroots organizations which attempt to achieve sustainable communities by defying the dominant American social movement that fails to consider minority political needs. As argued, this is significant because their combined mobilization or alliance organization constructs a new order of environmental democracy that is part of a possible fourth wave of American environmentalism. The only organizations that can effectively achieve the elaborate outline of a sustainable development and combat the present economically focused politics of technocracy are organized social justice movements—nongovernmental organizations comprised of collections of grassroots people united under common design and interest in changing the present situation. Simple united community activism brings new actors to undemocratic public policy affairs. This new social network is formed from labour factions, cooperatives, mothers' clubs, neighbourhood action federations, tribal groups, and religious unions.

As discussed, all of these groups embody the ideal of sustainable development and could possess enough energy to reform or change the current system of administered state-centred politics. Radical activists may not understand the genetic modifications of rBST or what the dangers of modern agrarianism are like a mainstream environmentalist expert would; however, if they communicate with or appreciate the political movements of the conservative environmentalists, they will organize with the sophistication of mainstream environmentalists. Unlike state and mainstream organizations, local social movements understand that a sustainable environment needs to have economic prosperity without sacrificing their health or the prospects for the health of future generations. Furthermore, social movements work more rapidly and are more flexible than other organizations due to their locality and accessibility of knowledge. In the discernable future, grassroots organizations devoted to achieving environmental justice can fundamentally alter the accepted political landscape and have a great impact on the outline of environmental democracy.

As the oldest environmental interest group in the United States, the Sierra Club has long been instrumental in passing legislation that benefits the environment. Nonetheless, it was not until recently that a formal environmental justice policy program was developed by the Sierra Club to better incorporate environmental justice goals into its agenda. According to many radical activists, valuing human diversity will maximize the Sierra Club's clout, publicly and politically. A Sierra Club official states: "Just imagine millions of people of all colours and backgrounds . . . [having] a real stake in the protecting an all-encompassing environment, from our public lands to our urban centers" (Baxter, 1997a, 102). If mainstream organizations continue outreach to radical green constituencies, solutions to the most damning environmental problems may be developed.

While still a mainstream interest group, the Sierra Club is beginning to look beyond Congress, state legislatures, and courtrooms to the special interest groups of grassroots activism. This is best declared by a Sierra Club publication that states, In a time when a citizen equals a consumer and when the public interests are eclipsed by special interests, the Sierra Club is seeking some changes (Cox, 1996).The Sierra Club believes that strengthening its democratic life is the best strategy for protecting the wild places of nature and the health of our communities. While the traditional assumptions made by established political actors fit poorly with passionate community activists, radicals seem to be learning from mainstream actors, and that is discernable in the fact that instances of great public damage by radical protesters are becoming less frequent.

Both groups have begun to move toward a process of respect rather than competition in environmental politics. As such, it is possible those sustainable justice organizations will serve as building blocks for a future, higher form of environmental consciousness and activity. This environmental grassroots movement could significantly alter the current dynamic of growth-oriented consumption capitalism which forces over-accumulation and reproduction with little to no regard for finite resources. Furthermore, modern society's undemocratic environmentalism could be dismissed as destructive with the praxis of policy interest pressure and common environmental issues. This is best illustrated by the use of Internet politics by radical and mainstream environmental organizations. That is, fringe groups that usually refrain from doing national campaigns and large organizational projects are using the Internet to work with the sophistication of mainstream organizations. A journalist explains this new phenomenon for radical groups:

> Even so-called ecoterrorists who have uprooted scores of university test plots across the country in the past year use the Net to organize their lawbreaking protests. In an Internet posting from Santa Cruz last week, Earth First! beacons, "You're all invited to sunny California for a weekend of workshops, training and fun! We also have plenty of [genetically engineered] crops waiting for your night time gardening efforts." Says Carl Pope, the Sierra Club's executive director: "I've never seen an issue go so quickly (Roosevelt, 2000)."

Motivated by protest in Europe and the proposed FDA regulations that would have allowed food certified as "organic" to contain genetically modified ingredients, most mainstream organizations are in agreement with radical groups with regard to food biotechnology. More specifically, most groups are petitioning the FDA to change its 1992 policy that stipulates biotech ingredients do not materially alter food and therefore do not need labelling and pre-market safety testing because biotech additives are no different than natural additives.

The most commonly cited and thoroughly developed mainstream environmental public policy solution for the existing ecological crisis is "sustainable development," or simply "sustainability." Sustainable development is the notion that societal development should meet "the needs and aspirations of the present without compromising the ability of future generations to meet their own needs"(Rifkin & Rifkin, 1992, 102). Figuring out how governments, industry, commerce, and communities can embody the principles of sustainable development is painstakingly difficult given the contemporary political commitment to economic growth.[33] That is, in an environmentally sustainable society, ecological conservation and economic development often seem contradictory. Among radical environmental activists, the dominant ecological theme is "environmental justice," a political practice that affirms the sacredness of the earth, ecological unity through human diversity, the interdependence of all species, and the right to be free from ecological destruction (Seager, 1993). According to advocates for environmental justice, the prescribed resolution is a progressive democratic association of groups with a strong ecological conscience. The goal of environmental justice, however, is very ambitious. It has never been practiced and can only be conceived through a patchwork of economic, gender, social, and racial equality.

The aim of finding similarities between the mainstream theory of sustainable development and the radical practice of environmental justice is to develop a stronger way to communicate to the public the real ecological risks of the contemporary age (Dobson, 1999b; Luke, 1999; Thiele, 1999). Although "sustainable development" is not in the language of justice theorists, and proponents of sustainable development do not talk of "environmental justice," I think that their belief systems can be closely associated. Leslie

[33] The World Conservation Strategy details a prescription for sustainable development by stating the following: an administrative system that is flexible and has the capacity for self-correction; an international system that fosters sustainable patterns of trade and finance; a technological system that can search continuously for new solutions; a production system that respects the obligation to preserve the ecological basis of development; a social system that provides for solutions for the tensions arising from disharmonious development; an economic system that is able to generate surpluses and technical knowledge on a self-reliant and self-sustained basis; and a political system that secures effective citizen participation in decisionmaking.

Paul Thiele espouses the same point of view in *Environmentalism for a New Millennium* (2000), where he promotes a "fourth wave" of the political movement for environmental policy that is without enemies, one that welcomes all types of ecological allies, unlike the three previous political waves (Thiele, 1999, 31). Thiele refers to the commingling of corporations in the Business Council for Sustainable Development and the preservationist radicals of Earth First as expanding our thinking and discourse about environmentalism.

More than expanding the environmental movement, I am interested in making its communication of environmental risks more inclusive and deliberative. In exploring a possible union of the often-paradoxical sustainable development notion and the seemingly utopian environmental justice movement, some authors discuss a combined vision that is more practical and agreeable than either concept is on its own. For instance, Timothy Luke argues that radical ecologists must force mainstream thinkers to express themselves less conservatively. Rather than using the scientific language of elites that serves the interests of the economics of the "green governmentality," Luke wants real-world discourse about dangers, threats, and risks. According to Luke in *Capitalism, Democracy, and Ecology* (2001), sustainable development talk must use the words of environmental justice to overcome the forces of traditional politics and properly address the global problematique—the emergent interplay of a number of interdependent problems in the world which endanger the planet.

What is important in Luke's work is the notion that sustainable development and environmental justice are not easily combined and require close analytical attention to develop a combined discourse. Similarly, Andrew Dobson's *Justice and the Environment* (2001) "recognizes that social justice and sustainability are likely to remain important goals for the foreseeable future and approves of both. His intent is to identify both synergies and conflicts across these two goals" for possible communicative solutions to ecological problems (Dryzek, 2000, 171). Developing the political theory and environmental thinking of risk communication, therefore, is the topic of the next chapter. The political theory of risk communication is very important because it serves as the basis for rational, administrative, and scientific practices of both risk assessment and management.

Environmental assessors and managers must communicate their elite technological analyses and environmental regulations to the general public because whether experts realize it or not their communication and therefore risk judgments are shaped by social events, values, and actions. Unlike most risk scholars who argue for a risk communication that privileges consumer risk by describing the erroneous tendencies of producer risk, or who pursue a strategy of risk communication that favours elite judgments rather than lay perceptions of risk, my research hopes to inform both sides of the risk message (K. Shrader-Frechette, 1989). In the following two chapters, I achieve this goal by explaining the underpinning communicative action of critical social theory and discuss the pragmatic limits both experts and the public have with regard to risk messages, transmissions, collaborations, and facilitations. Expanding the liberal understanding of administrative practice of communication may minimize the difficulties of democracy in the technocratic age.

CHAPTER SIX

THE RISKS OF ENVIRONMENTAL DEMOCRACY

1.INTRODUCTION

The past twenty years has seen the publication of more than 1,800 research papers on the only genetically engineered veterinary drug or growth hormone used to increase the production of milk in dairy cows, rBST (Powell & Leiss, 1997, 124).[34] Regulatory agencies in at least thirty-one countries have publicly announced that all foods produced from animals with rBST technology are safe for human consumption (P. Thompson, 1998b). Breaking with its departmental policy of keeping internal decisionmaking processes confidential, the United States' FDA explained in an academic journal the methodology it used to conclude that milk from rBST cows is as safe as and no different than milk from cows not treated with the growth hormone (Juskevich & Guyer, 1990). The Department of Health and Human Services (DHHS) released a press statement—on November 5, 1993, nine years after the initial application by the St. Louis based Monsanto Corporation—explaining that even though rBST had no human health or environmental risks, there was a moratorium on sales until Congress could consider all impacts of the new biotechnology (Powell & Leiss, 1997). In additional to all of these communications, there was extensive media coverage of the first major product of agricultural biotechnology to move from the laboratory to the farm.

Together the published articles explaining the mathematical risk assessment of rBST, the press release from DHHS, and various announcements from multiple agencies that insisted recombinant bovine growth hormone was safe, are among the most innovative attempts of risk communication in American environmental history. Regardless of all of these transfers of information and the successful use of rBST in the United States since February 2, 1994, many Americans question the authority of experts and public officials who have and continue to make technological

[34] For an annotated bibliographical listing of 657 citations of rBST please see United States Department of Agriculture, National Agricultural Library, Quick Reference Series, 1994. "BST-Bovine Growth Hormone January 1987-January 1992 and January 1991-December 1993." Washington, D.C.: USDA Press.

decisions about food biotechnology (Deanna Grobe & Douthitt, 1999). Notwithstanding industry predictions of everything from increased nutritional benefits to lower prices for consumers, many American consumers remain suspicious about the risks of food biotechnology in general and rBST in particular.

Genetic engineering, especially with food systems, replaced nuclear energy in popular culture as the primary example of modern science and technology threatening human autonomy and trust in government (Feenberg, 1999; Krimsky, 1991; Powell & Leiss, 1997). Recent commentaries suggest that, unbeknownst to the average consumer, over two-thirds of food bought by the average American—from milk to soybeans—contain genetically engineered components (Burros, 1999b). Just as the environmental scholarship of twenty years ago disputed and critiqued the technical rationalism of the nuclear industry, many environmentalists today are decrying the technical rationalism of genetically industrialized agriculture. In order to move beyond philosophical critiques of rationalism and create policy processes that have learned from the unresolved politics of the nuclear age, theorists have formed the idea of environmental democracy, the development of a bureaucracy and citizenry that is able to deliberate adequately about the human health and ecological challenges presented by society's technological advances.

Among the more precise articulations regarding the difficulties of practicing democracy within the economic and social norms of modern environmental policy is the following:

> Expertise legitimates power in society at large, and "citizenship" consists in the recognition of its claims and conscientious performance in mindless subordinate roles. The public sphere withers; a literal reign of silence is instituted as one-way communication replaces dialogue and debate throughout society (Feenberg, 1999, 101).

Analysing the unwanted consequences of a technocratic society on the human condition has been one of the central tasks of the philosophy of critical social theory.[35] According to this school of thought, the fundamental problem of democracy today is the potential loss of agency in a culture that technologically overpowers independence, autonomy and self-government. Three concepts that can help reduce the deterioration of autonomy, that will

[35] Jurgen Habermas in *Philosophical Discourses of Modernity* (1987) states, "Critical Theory was initially developed in Horkheimer's circle to think through political disappointments at the absence of revolution in the West, the development of Stalinism in Soviet Russia, and the victory of fascism in Germany (p. 116)." It should be noted that Habermas disagrees with the first generation of the Frankfort Circle that believed that reason was inextricably intertwined with domination. Habermas maintains that the reason is the solution for some of the problems of modernity, most specifically, alienation. Habermas believes that the full potential of modern reason is found in moral-practical speech and in differentiating the three types of human rationality (the cognitive-instrumental, the moral-practical, and the aesthetic-expressive.)

be examined in the balance of this chapter, are the following: (1) Habermas's "technization of the lifeworld," (2) Herbert Marcuse's "one-dimensionality," and (3) Ulrich Beck's' "risk society." Given that the overall project of critical social theory is emancipation, the ideas of these three philosophers can be read as an attempt to construct individuals and a society that continually question technology, functionally learn within it, and ultimately manage technology.

The foundational theorist for attempting to democratise environmental politics and processes is Jurgen Habermas (Brulle, 2000; Hajer, 1995; Williams & Matheney, 1995). It has been argued that Habermas's communicative ethics could be used to transform the traditional political understanding of the good life. Through a deliberative learning process of language between citizens and experts, a practical vision of the abstract notion of environmental democracy is found in Habermas's suggestion of a communicative administration. In fact, Habermas's prescription of an open political administration has been acknowledged as the theoretical justification for laws that oblige governmental agencies to publicly release information regarding their actions (Feenberg, 1999). Like other commentators, however, I assert that such information transfers fall short of Habermas's ideal of deliberative participation. Quite simply, transparency of public bureaucracy is not the fruition of Habermas's notion of public deliberation. In this chapter, I claim that risk communication concerning rBST represents an imperfect or impoverished version of Habermas's communicative action.

Habermas's full understanding of a democratised public administration is best seen through his early works, which emphasize communication, rather than his most recent texts that stress law. Habermas's inclusive bureaucracy is elite driven, and demands an engaged, responsible, and organized citizenry. In order to make a significant step toward democratizing a modern technically based society, the works of Marcuse and Beck will be utilized to make Habermas's ideal liberalism a real liberalism. With the help of other theorists a fuller practice of Habermas's communicative ethics can be achieved to yield a better process of risk communication for American environmental policy, generally, and a more democratic understanding of public policy debates about food biotechnology, particularly. Constructing a political space that discursively communicates expert interpretations of harm and public perception of risk is achieved by augmenting Jurgen Habermas's rational liberalism with Herbert Marcuse's writings on a cultural revolution of political thinking among the masses and Ulrich Beck's work on encouraging a new social cohesion and political action of the citizenry that is spurred by humanity's shared risk. Habermas's theorizing of a participatory administration that is open to influence from public inputs supplements many of the main tenets of critical social theory.

In his 1996 work, *Between Facts and Norms*, Habermas describes his proposal to make public policymaking more deliberative:

> Insofar as the administration cannot refrain from appealing to normative reasons when it implements open legal programs, it should be able to carry out these steps of administrative law making in the form of communication and according to procedures that satisfy the conditions of constitutional legitimacy. This implies a "democratization" of the administration that, going beyond special obligations to provide information, would supplement parliamentary and judicial controls on administration from within (Habermas, 1996, 440).

Making political institutions function beyond mere representational mechanisms—such as putting them under public control, into the public sphere or within public reflection—is a common theme for critical social theory. Habermas challenges both the current structure of policy deliberation and mainstream liberalism when he recommends that regulators do more than publicly release legal decisions.

Food biotechnology is making direct genetic modifications of microbe, plant and animal DNA in order to increase profit, improve breeding, or intensify traits desired by consumers. As these new products are successfully making their way through the federal regulatory system to the grocery store, concerns are being raised by specific consumer groups and ecological scientists, but the potential environmental threats of this new agriculture are not being discussed by the majority of American citizens.[36] The most significant factor preventing such a public conversation is that governmental food safety agencies and other scientific regulators of biotechnology often only communicate with each other in an elite language for institutional participation. When such institutional communication is made public, it is for the purpose of announcing already decided risk calculations, not to engage the citizenry in a deliberative participation of risk analyses. Such an information transfer eliminates the possibility of Americans discussing the consequences of living at risk.

Risk analysis is the bureaucratic process of making societal choices about ecological dangers and human health hazards by determining the probability of harm using scientific techniques of evaluation. Within this technical analysis of safety, the notion of a politically debated "acceptable risk" is developed by risk administrators but never adequately communicated to the public. For instance, the majority of citizens in the United States seem to

[36] A recent study by the International Food Information Council found that 71 percent of consumers in the United States rated themselves poorly informed about food biotechnology. And, most consumers do not know that many environmentalists are questioning the regulatory policies regarding food biotechnology.

trust governmental and industrial scientists to guard against the risks from modern agriculture by attending to both the safety of food produced and the integrity of the environment used to produce it. As revealed in recent public opinion polls and acts of environmental civil disobedience against biotechnology research and sales, however, some consumers are beginning to question their trust in elites and react to their lack of communication by governmental administrators and industry technicians (Roosevelt, 2000). The failure of rational experts to converse intelligibly with society about risk eventually can result in massive political distrust of their ability to assure public health, safety and environmental protection (Kasperson, Golding, & Tuler, 1992).

Both experts and citizens incorrectly perceive environmental risk analyses—such as those involved in the debates over food biotechnology—as technically rational and socially apolitical. In fact, such environmental risk analyses are based, at least implicitly, on particular liberal political theories. Yet, risk practices are muddled and misdirected because risk practitioners fail to realize completely the systematic liberal philosophies that undergird their work. Making theoretical foundations explicit, however, should lead to improved practice. Therefore, we may construct a better political life by examining critical social theorists who argue that a richer public discourse will be created by including competing values in decisionmaking. Rather than a transmission of data, risk communication can be a process of building consensus of interests among technical experts, policymakers, and affected members of the public. If risk communication used some of the philosophical insights of Habermas, Marcuse, and Beck it could be realized as a constructive discourse instead of a one-way, often incomprehensible, information transfer from experts to lay people. Such an ideal practice of liberalism could be a theoretical resolution to the problems of human agency in the technological society.

2.HABERMAS'S IMPORT

Habermas once said that environmental problems are initially understood as abstract challenges that "call for technical and economic solutions, which must in turn be globally planned and implemented by administrative means" but eventually "affect the organic foundations of the lifeworld and make us drastically aware of the standards of livability"(Habermas, 1987, 394). Students of Habermas have combined such ecological thoughts with his theory of discourse ethics to organize a notion of deliberative democracy that could be directly applicable to environmental problems and engineered toward improving risk communication.[37] The political import of Habermas's communicative philosophy is best explained as follows:

> Deliberative democracy promises legitimate—that is, morally justifiable and rationally produced—solutions to vexing political problems. Especially when these problems are difficult, affording no clear way to arrive at unequivocally satisfactory solutions, deliberation recommends itself because it relies on a broad consideration of alternative solutions, increasing the likelihood that the perspective held by all members of a heterogeneous community will be given a voice (Sanders, 1997, 347).

Deliberative—also called discursive—democracy focuses on solving problems through having participants exchange diverse ideas, perspectives, and criticisms. To some degree, this is the type of mainstream-radical discourse talked of in the previous chapter that is emerging in American environmentalism. The goal of such progressive exchanges is to try to articulate the common good; according to Habermas, this ambitious objective can be produced through the communicative action of the ideal speech situation (Dryzek, 1990, 1996a).

Because deliberative processes recognize all positions as valuable, discursive politics should increase the ability of experts to consider multiple values in decisionmaking and the opportunity of citizens to contribute to the discourse. Indeed, such an approach would aid the technical language of rational calculus or risk assessment, which fails to capture many citizen concerns about the environment. This advance in decisionmaking should improve environmental policy, because practicing Habermas's open administration requires that complex issues be understood, evaluated, and prioritised as a democratic community (R. Bartlett & Baber, 1999). Unlike traditional risk communication, which elevates human interests above all others, discursive democracy may consider both anthropocentric concerns for human health and ecocentric hazards to environmental sustainability (Mathews, 1996). Such an expansion allows administrations to enhance the potential of democratic policy processes because such a perspective emphasizes the ability of rational experts to critique themselves positively and come to new, more inclusive risk communications.

Habermas's project is the construction of a political arrangement based on the standard of a rational society. His social theory envisions the establishment of a community with unhindered rational discourse as the foundation for a legitimate political order. Although Habermas has been ideologically characterized as everything from a neo-Marxist to a new conservative, his objective has always been to develop "a theory of rationalization that requires the elaboration of categories and concepts for a systematic examination of the character and different modes of rationality

[37] Although interpreted as an exceptional theorist for environmental politics by many secondary authors, Habermas himself has said very little about the green movement himself.

and an explanation of how they are concretely embodied in a social and cultural life"(R. J. Bernstein, 1985, 4). Habermas's commitment to developing a universal perspective on rationality for politics is first noticed—and perhaps most lucidly written—in his earliest works. *Toward a Rational Society* in 1968, *Theory and Practice* in 1970, and *Legitimation Crisis* in 1973 offer critiques of cultural norms, knowledge and behaviours in order to insure the survival of human agency in the technological era of advanced capitalism.

Even though Habermas prescribes a structured rationalism for political and social life, it is dissimilar to the present and traditional rationalism. Habermas describes the latter rationalism as "growth of productive forces; extension of power of technical control" and the former rationalization as "emancipation, individualization; extension of communication free of domination"(Habermas, 1968, 93). Habermas has developed a theory for rerationalizing communication—the transformation of contemporary society through self-emancipation—with the emergence of free and unconstrained discourse (Rehg, 1994). That is, Habermas expects reasonable human agents to be capable of identifying subjective or personal rationality (the cognitive-instrumental, the moral-practical and the aesthetic-expressive) from interpersonal dialogue in order to construct an objective or social rationality. Habermas's theory of communicative competence or "universal pragmatics" maintains that implicit within all human communication is a presupposition of an "ideal speech situation." The ideal speech situation is the normative structure in rational discourse for defining the ideal conditions of political arrangements and uniting practical and theoretical knowledge claims (Rehg, 1994).

Habermas's complete project, from the critique of contemporary science to the reconstruction of a new social rationalism, is found in perfecting human communication beyond linguistic interpretation but not reducing it to an empirically analytical science (McCarthy, 1978). Habermas explains his theory of communicative action or universal pragmatics by writing:

> Up to now we have conceived of action in terms of dealing with situations. The concept of communicative action singles out above all two aspects of this situation management: the teleological aspect of realizing one's aims and the communicative aspect of interpreting a situation and arriving at some agreement. In communicative action participants pursue their plans cooperatively on the basis of a shared definition of the situation (Habermas, 1987, 126).

Habermas is attempting to establish a rerationalization of politics and ethics with a process of discourse that attempts to achieve consensus among different members of a community. Although the details of Habermas's theory of communicative action are perplexing in their complete linguistic structure, the leading ideas for constructing a deliberative democracy and needing a communicative democracy can be sketched briefly.

The starting point for communicative action is an analysis of the universal pragmatics of all discourse, where the speaker and listener are oriented to the establishment of mutual and reciprocal understanding with the objective of finding a truth on which they can agree (Habermas, 1979, 3). For Habermas, all communicative action takes place against a background consensus. Consensus for Habermas is based upon "comprehensibility" among all ongoing communicative acts. Ideally, the only force that should prevail in such a discursive consensus is the "force of the better argument." Habermas is, of course, aware that he is describing an "ideal type of dialogue" that fails to exist in most policy conversations, where individuals strategically seek to manipulate one another (Stone, 1997). Nonetheless, Habermas's major point is that in attempting to achieve the ideal communication or in resolving breakdowns in critical communication, there are universal validity claims "set in the general structures of possible communication" for "the intersubjective structures of social reproduction" (R. J. Bernstein, 1985, 19). That is, because Habermas believes that the normative basis of critical knowledge is rooted in the structure of every action of communication, the only means of attempting to establish consensus after a breakdown is through more deliberative communication or discourse.

Inherent in all discourse is a common idea of a possible situation where consensus is ideally re-established solely upon the basis of the best, most rational argument among the competing options for the resolution of the point of disturbance. This inherent situation which underlies all acts of communication is Habermas's "ideal speech situation." The ideal speech situation is a condition of discourse free from domination, with equality of participants and the exclusion of distortions of self-perception. The goal of Habermas's rational discourse is mutual trust and reciprocal understanding. The principal link to the ideal speech situation, which emphasizes the practical rather than theoretical nature of communicative competence, is again Habermas's notion of rationality. For Habermas, a consensus on truth is not a mere agreement or compromise of discourse; rather, a consensus on truth is only achieved in a trusting process of rationality (Habermas, 1979, 117). By rationalization, Habermas means a process of increasing the trust and reasonableness in social action.

Risk communication, by both conventional and symbolic understanding, is a practice of Habermas's theory of communicative action. Theoretically, the conventional meaning of risk communication is the transmission of technical or scientific information from elites to the general public, and the symbolic understanding is any public or private discourse that "informs individuals about the existence, nature, form, severity or acceptability of risk"(Krimsky & Plough, 1988, 5). Because risk events are composed partly of physical processes and partly of socially constructed phenomenon, all risk decisions must be deliberately examined in their political context to obtain a full and truthful accounting of the factors involved (Krimsky & Plough, 1988). As a consequence, it is only through a deliberative communicative action of democracy that the complex socio-political perceptions of risk can be understood and managed. Furthermore, it is in this process of deliberative communication that previously considered "irrational" citizens become reasonable for Habermas.

The purpose for rational discourse or deliberative argumentation is to resolve disturbances in the act of ideal communication. The ideal speech situation then becomes the foundation of ethics and politics and the benchmark of both practical and technical discourse claims. According to many commentators, for public policy conversations, this means the following (Bohman, 1996; Dryzek, 1996a; Fisher & Forester, 1993):

> Habermas' insistence on the force of normative concepts as inherent in everyday language does not lead him to reject compromise as a means of political consensus formation; rather, it moves him toward a model of decision-making in which the pragmatic give-and-take of daily politics is guided by any implicit understanding of moral values that are present even when they are not in the foreground. Clearly, Habermas thereby rejects a reading of communicative action that stresses the split between an idealized speech situation on the one hand and, on the other, the rough-and-tumble of the "real world" where such normative considerations do not affect the decision-making process (Habermas, 1994, xxiv).

The ideal speech situation is more than a mere theory of communication in a linguistic setting with implications for a theory of truth. Its epistemological foundation for the determination of facts and values becomes a practical hypothesis for deliberative pubic policy conversations.

The final goal of Habermas's 1987 *The Theory of Communicative Action*—where much of his thoughts on discourse ethics are found—is to demonstrate the successful nexus between political practice and a theory of socialization (Rehg, 1994). As he states, "we cannot simply describe, explain, and understand social forms of life without explicitly or implicitly evaluating the rationality of social action and action systems"(Habermas, 1987, 276). Communicative action has great concern for reasonable deliberation because it stems from the fundamental reflection upon what it truly means to be human. "The call for more deliberation is, I argue, a demand for a more rational political order in which decisionmaking at least involves the use of public reason"(Bohman, 1996, 2). Habermas's normative dialogue with a participatory policy design promises a fundamental solution to the conflict between the lay citizen and the rational expert in the modern technocratic society (Feenberg, 1999, 125).

In short, communicative competence is a politics that attempts to make a society civilized. Habermas's theory of rationality can be used as a defence of human autonomy during the modern age of science or the "technization of the lifeworld" by affirming that we all must learn from communicating with each other in society (Habermas, 1987, 400). Habermas optimistically believes that the human condition shows a design of political enlightenment in social communication. Before Habermas's universal pragmatics, communicative competence, and the ideal speech situation can be achieved, the starting point must be an analysis of the pragmatics of speech whereby the speaker and the listener are oriented to an establishment of mutual trust and reciprocal understanding with the object of finding a rational consensus. Habermas painfully emphasizes that all viable communication necessarily depends upon the assumed consensus of the "validity claims of comprehensibility, truthfulness, and rightness"(Habermas, 1979, 3). If the delivery of understanding in communicative competence is disturbed in some manner, the appropriate solution is to improve the nature of discourse.

The serviceability of Habermas's deliberative democracy in American environmental politics is found in questioning the following: what do participants in risk assessments say, how are their perceptions and communications of risk evidence relevant, and why does their discourse influence policy decisions? In responding to these questions, many scholars maintain that there are "hidden hazards" of environmental risk talk that are thoroughly implanted in our belief systems, scientific practices, and social relations (Kasperson et al., 1992). For instance, during the research protocols of rBST, the FDA's Center for Veterinary Medicine and the General Accounting Office (GAO) reported that cows injected with rBST once every fourteen or twenty-eight days will increase their milk production by fifteen to twenty percent, with no significant side effects to the animal that could not be treated with antibiotics (GAO, 1992). The truthful hazards of animal well-being were hidden in the evidence used by policy experts, as is evidenced by the Monsanto package insert, which lists the risks to cows. It reads:

> . . .increased cystic ovaries and disorders of the uterus; higher incidence of restrained placenta; increased risk of clinical and subclinical mastitis; increased digestive disorders such as indigestion, bloat and diarrhea; increased numbers of enlarged hocks and lesions of the knee; disorders of the foot; and injection site lesions which may remain permanent (Mepham, 1996, 108-109).

Uncovering these hidden risks can only be achieved by a democratic decisionmaking among natural science, social science, and public communities.

The attractiveness of agricultural biotechnology to farmers is an important consideration in the decision to accept rBST; but, that appeal can only be genuine if the communication of risk is truthful and complete. Honestly, the only way to achieve this type of risk communication is to debate subjective perceptions of risk without violations of expectations that people may have in social relations (Kasperson et al., 1992). If all of the relevant voices affected by the proposed environmental action are allowed to debate in a truthful way, this could mean the democratic evolution of a risk communication by regulatory bodies.

The flaw in the underpinning of rational decisionmaking is in not having a discursive framework for questioning the authority, infallibility, and completeness of science in policymaking. Habermas's model of decisionmaking would consider values in technological decisions characterized by the free deliberations of collective needs without hierarchy or formal rules. This type of analysis will improve political dialogue because it considers a wider range of other alternatives and institutes more communicative attempts to resolve or ameliorate social problems. Habermas's significant import to risk communication is that critical thinking concepts are most applicable during the policy formulation or agenda setting stage (Ingram & Rathgeb, 1995). Critical theory focuses inquiry and shapes attention among policy planners at the genesis of policy argumentation because the planner is not a professor of facts but an organizer of information (Fisher & Forester, 1993).

As previously stated, however, there are some problems with the marriage of Habermas's theory of communicative action and the practice of American environmental politics that are not resolved by Habermas's theory of discursive democracy. The first, and most damning, is that Habermas's rationality (with the emphasis on cognitive-instrumental and moral-practical over aesthetic-expressive) is only discerned in human speech, thus the moral integrity of the nonhuman world could be threatened by traditional political actors who are engaging in discourse ethics but are not environmentally conscious. In *Environmentalism and Political Theory,* it has been written:

> According to Habermas' schema, a norm is considered "right" if it is achieved via a consensus reached between truthful, uncoerced, and rational human agents. It follows that if a "speech community" agrees, after free and rational discussion, to direct technology in such a way as to continue to manipulate and subjugate "external nature," then Critical Theory can raise no objection since its concept of emancipation has been exhausted (its exclusive concern being with human self-determination) (Eckersley, 1992, 110).

According to some environmental scholars, Habermas's open administration incorrectly assumes that policy participants will be anchored in a sustainable worldview. That is, Habermas's procedures of democratic discourse will have the political legitimacy to exploit the environment, as well as rerationalize politics to value the environment.

While an important critique, I think this concern can be largely resolved by assuring that the participants who engage governmental and industry experts include citizens with a wide range of perspectives, including some who have not traditionally participated in the political dialogue. Communicative action, like all political discourses, is intentional. Its goal is to reveal facts and values from multiple points of view. Untraditional political actors who would change the traditionally rational discourse, however, will by their very orientation acknowledge this point. Marcuse's revolution of political participation for citizens who act according to inner drives, direct-action, or non-rational motives best explains the importance of expanding risk communication to non-traditional political members.

The second and related criticism of Habermas's inspired discursive democracy is that even if, as Marcuse argues, citizens can be inspired to use their aesthetic-expressive sphere of reason, they might not be able to articulate the rights, autonomy, and integrity of the natural world forcefully in a policy discourse. To address this problem I draw on the critical social theory of Ulrich Beck and the notion of communicative democracy. According to Beck, once individuals acknowledge that their perceptions of risk are valid political positions, they will begin to advocate a modernity that is ecologically rational. The notion of a communicative democracy is presented as a subtly nuanced variation of the notion of discursive democracy, which may tend to privilege elites. Such a theoretical augmentation of Habermas's democratic dilemmas helps acknowledge the real liberal potential of risk communication.

3.MARCUSE'S PROJECT

Herbert Marcuse's lessons for aiding Habermas's notion of a deliberative risk communication are threefold: (1) there will always be a discord between scientific experts and the public around issues of environmental and health risks; (2) although it is true those who control the discourse on risk will most likely control the political battles, the success of risk communication should not be measured by the degree to which popular attitudes reflect the technical rationality of risk (Plough & Krimsky, 1987); and (3) the real goal of risk communication should not be building societal confidence about the safety of a contested technological product, but should solidify trust in the process of approval by having members of the society meaningfully participate. That is, the cultural perception of a risk event will never be identical with the scientific understanding of that same event, but they do not have to be inconsistent because they are communicable. A reasonable decisionmaking process that incorporates some of the Marcusian "irrationalism" of the citizenry may yield better decisions because it will create a political culture that will be less repressive for all.[38]

Reacting against the mainstream ideas of idealism, positivism, and scientific intellectualism, Herbert Marcuse did not emphasize philosophical intellect or rational behaviour, but instead emphasized direct action and inner drives. Marcuse wanted an innovative approach to understanding the human condition in which old certainties, such as the role of intuition and experience, would provoke individual political consciousness in a collective culture of repressive malaise. *One-Dimensional Man* (1964), Marcuse's most celebrated work, is a treatise on the uncritical and conformist technologized politics of industrialized Western society, capitalist and communist alike. Novel insights about the difficulties of functioning only in the institutionalised rationality of liberal politics are also found in his 1966 *Eros and Civilization* and 1972 *Essay on Liberation.*

In these works Marcuse formulates many philosophical moves that can be helpful in creating a more participatory interaction between the mass citizenry and the technical elite (Marcuse, 1972). Marcuse's understanding of the cultural tension between the past and present is the foundational block for creating a new political education for self-governed citizens who will redefine today's decadent political society controlled by elites. In Marcuse's terms, there is a need to create a "new science" grounded in non-instrumental rationality in order to break the voluntary and self-administered repression of contemporary politics and economics (Kellner, 1991). Marcuse's writings illuminate a political culture necessary for bringing to fruition a theory of risk communication that could exercise the reasonable practice of environmental democracy.

For Marcuse there is an eternal conflict between the theoretical and tragic worldviews (Biro, 2000). In contrast to the rationalism and optimism of the theoretical worldview, the tragic worldview esteems pleasure as the foundation of culture and society. The theoretical view deludes humanity into believing that earthly happiness is possible for all, while the tragic or practical view accepts "that this freedom and satisfaction are transforming the earth into hell" (Marcuse, 1966, xiii). Although there can be no final redemption from the suffering of life and history, a "nonrepressive" culture can create the conditions for a "worldly asceticism" and "pleasure of existence" (Marcuse, 1972). This is unfortunate because, Marcuse explains, "Today, this union of freedom and servitude has become natural and a

[38] Please note that I am not attempting to distort Marcuse's mature philosophical formulas about developing new political sensibilities into the prevailing social order of American liberalism. That is, I am not painting Marcuse as a democratic reformer. Instead, I am using him to add colour, life and resonance to the public citizenry needed in Habermas's critical theory.

vehicle of progress" (Marcuse, 1966, xiv). Marcuse believes the natural and the human world await liberation of oppression and pain (Marcuse, 1966). Unlike many thinkers who attempt to reform politics with some theory of ideal participatory liberalism, Marcuse hopes to describe the tragic political environment of one-dimensional liberalism in order for the citizenry to promote a real politics of revolution (Biro, 2000).

Marcuse holds that contemporary politics rests on a certain faith in a destructive rationality. Living in this uncritical, unexamined, and unreflexive modern culture is what Marcuse identifies as one-dimensionality, defined as follows:

> . . .thought and behavior in which ideas, aspirations, and objectives that, by their content, transcend the established universe of discourse and action are either repelled or reduced to the terms of this universe. They are redefined by the rationality of the given system and of its quantitative extension (Marcuse, 1966, 12).

If we are to properly appreciate the meaning of culture and participate in an authentically political life, for Marcuse, it is necessary to reassess the value of science and knowledge. The transfiguration of nature will occur with a cultural liberation that allows artistic style and expression in political life. Marcuse argues that the goal of humanity cannot lie in its historical end, but only in its ahistorical transfiguring. Marcuse attacks the modern political culture for burying the spirit of the youth with the weight of history. According to Marcuse, history might be serviceable but it is destructive when used for the possession of knowledge for traditional rationality.

A political system must inform the citizenry that the ahistorical and the historical are required in equal measure for the health of an individual and a society. More importantly, this same dichotomy is true for science and society. Marcuse writes:

> To formulate the relation between science and society in this way is assuming two separate realms and events that meet each other, namely, (1) science and scientific thought, with their internal concepts and internal truth, and (2) the use and application of science in social reality. In other words, no matter how close the connection between the two developments may be, they do not imply and define each other. Pure science is not applied science, it retains its identity and validity apart from its utilization (Marcuse, 1964, 154).

To be an authentic human being is to live within this tension between the historical, scientific, or rational and the ahistorical, political, or irrational. The capacity to live ahistorically is, Marcuse contends, more vital and more fundamental than its counterpart, because "historical backwardness may again become the historical chance of turning the wheel of progress to another direction"(Marcuse, 1966, xvii). Citizens must seek to transcend the one-dimensional view of the technological society and affirm "self-mastery" or self-government as "the way of life" which is dissimilar to the traditional economic goal of society.

In an era ruled by rational self-interest, Marcuse wishes to regenerate social life with an emphasis on genuine political discipline that conceives the individual as part of an organic whole. Modern political culture rests on economic individualism, in which those virtues that cultivate self-government have been eroded. In the modern polity, the basis of the individual's obligation to society rests almost entirely on self-interested economic grounds. Thus, the political sentiments and virtues that Marcuse requires are absent in the modern world. The "ethical impulse," the "transvaluation of values," or the call to an "unrepressive society," for instance, are wrongly undermined and rendered improper to the character of modern liberal individualism, capitalism, and socialism (Marcuse, 1966).[39]

For Marcuse, the contrast between his views and the perceptions of contemporary society lies in resolving the cultural importance of work, commerce, and progress (Marcuse, 1966, 44). The modern world is characterized as an age of work in which the most esteemed societal goods are produced from the dignity of human labour. Marcuse, however, does not glorify labour through a work ethic because he knows that devoting a life to labour makes it impossible for a person to transcend the values of the one-dimensional society. Marcuse knows that he is saying something that runs counter to the cherished ideal of the industrialized state. For Marcuse, the origins of the state are violent and bloody, for it is in the nature of power that "Western civilization has always glorified the hero, the sacrifice of life for the city, the state, the nation; it has rarely asked the question of whether the established city, state, nation were worth the sacrifice" (Marcuse, 1966, xix). The state could, however, be regarded as the goal and the ultimate aim of the sacrifices and duties of the individual if the citizens could engage the bureaucratic experts in meaningful discourse about the construction of society. The duty of the modern citizen is not as industrialized worker but as political actor because he or she possesses relevant political knowledge to create a less repressive state and culture.

Because genuine political instinct and virtues are absent in the modern period, Marcuse discerns that instead of the state being a means to the production of culture, it is being reduced to a means for the furtherance of

[39] Marcuse, like many theorists frustrated by American interest-group liberalism, asserts that capitalism has led to remarkable economic benefits, but only at the cost of considerable experiences of social deficiencies. Similarly, democratic liberalism is bound up with industrial and commercial expansion; while this expansion is a product of individual initiative; it has created an oppressive society that favours technological exploitation over personal freedom (Marcuse, 1964, 2-3) And socialism, according to Marcuse, is dismissed because its assumed freedom actually results in acceptance of advanced industrial structures, norms and behaviours, and not social peacefulness.

the self-interested individual. According to Marcuse, the object of modern politics should not be to establish the state purely on the rational basis for the pursuit of economic ends. Rather, Marcuse wants to combine the erotic or aesthetic with the political or rational in the human condition.

Marcuse argues that while it is true that the elite control the political battles in modernity, the success of a culture should not be measured by the degree to which popular attitudes reflect the status quo. Instead the real goal of politics should be about solidifying trust in the process of decisionmaking by having all members of the society participate meaningfully. More importantly, this reform must be motivated by the citizens demanding liberation and not by the experts communicating freedom. According to Marcuse, in order to spur action for participation, rational self-interest should be dismissed for pleasure of self-government. Although he does not offer any new political structure, Marcuse suggests that a new political student of action should create new institutions. This Marcusian scholar is conscious of the battle for the survival of agency and views it as the only "truth-value" in politics. More importantly, this political truth will be most evident for citizens living with involuntary societal risk.

Although I am suspicious of theorists who reread Marcuse as a philosopher who prescribes sustainable development to a modern age that fails to recognize its ecological limits, I do maintain that his writings have relevance to environmentalism (Lukes, 1994). Moreover, I would not want environmental policy to be determined by an irrational biological necessity for the political fight for the theoretical worldview. Such a socially progressive notion would be unable to acknowledge the limits of understanding risk only through the perceptions of the public. Dismissing the scientific judgments of risk for the whims of the democratic polis is a precarious policy suggestion given the technical nature of most environmental issues that pose risks. I would assert, however, that inner drives, direct action, and reasonable experiences should be considered in addition to administered environmental public policy and rational risk analysis. Granted, I think it is philosophically naïve to think that Marcuse's citizen of "new sensibilities" will be "liberated from the work world's imposing upon him alien needs and alien possibilities. . .free to exert autonomy over a life that would be his own" (Marcuse, 1964, 2). Nonetheless, Marcuse's writings are serviceable for understanding critical social theories of deliberative and communicative democracy in order to create a culture that can fight the "historical alternative" to a political culture that has a "pacification of existence"(Marcuse, 1964, 17).

Regardless of the lack of any scientific evidence, environmentalists have continued to treat the FDA approval of rBST with forceful scepticism. This cynicism and uncertainty can be seen from boycotts on rBST exports, through policy demands for new milk labelling, to court proceedings petitioning for a moratorium on sales in certain states. The consuming public's disapproval of rBST is significant not only because the business of food production must consider the economic demands of the consuming public, but also because the government must consider the unscientific or "irrational" concerns of its subjects in order to be respected as a trustworthy institution. In fact, former Monsanto executives have stated that their decision that the first food biotechnology marketed be one that benefited specific farmers rather than general consumers was a strategic mistake (Eichenwald et al., 2001). Internal Monsanto memoranda that show some company executives suggesting that the company not continue scientific development on rBST because they feared that tinkering with a product consumed by children would ignite a national outcry best illustrates the problems of operating in one-dimensional rationality (Eichenwald et al., 2001).

It is the responsibility of government health agencies, scientists, and manufacturers of the products to provide early, adequate, and honest public education that is responsive to the public's concerns (Burton & McBride, 1989a). Attention to the concerns of the public might be the only means to prevent social distrust over this and future agricultural products of biotechnology and will, therefore, allow the public to form thoughtful assessments with respect to acceptance or rejection of each product (Burton & McBride, 1989a). The pursuit of a Marcusian goal, however, would not lead to public education and elite information sharing used to construct a more legitimate analysis of risk. Instead it would be in the creation of a policy framework that would capture society's fullest understanding of the values at risk. That is, a critical social theory of risk communication would be a decisionmaking process among expert technicians and informed citizens who would deliberate clear and timely explanations of the nature of the risks in a democratic society. This can only be done, however, with risk-conscious citizens.

4.BECK'S CONTRIBUTION

The critical social theory of Ulrich Beck also promotes a rerationalization of politics that could be used for constructing a better risk communication. Beck's notion of reflexive modernization, found in his 1992 Risk Society and 1995 Ecological Politics in an Age of Risk, recognizes and gives marginalized citizens of the polity Marcusian vitality to engage in Habermas's community discussions of risk. "Rather than seeing an orderly class hierarchy as the structure of social relationships, Beck sketches a disorderly and cross cutting set of relationships based on who is vulnerable to whom, in what respects and with what resources for reversing, revising and refocusing the risk relationship" (P. Thompson, 1998b, 237). Developing Marcuse's call for inner drives in political culture and evolving from Habermas's deliberative decisionmaking, Beck creates a philosophical notion of communicative democracy from the societal modernization of critically evaluating one's societal position of living at risk (Beck, 1992). Specifically, for Beck the project of traditional politics is opening itself because it is simultaneously challenged by globalisation, individualization, gender revolution, underemployment, and environmental risks.

Drawing upon the theory of deliberative democracy and the environmental practices of risk, scholars have developed the notion of a communicative democracy (Sanders, 1997). Like deliberative democracy, communicative democracy values discursive participation in public discussions. Communicative political theorists, however, assert that deliberation is not culturally neutral or universally applicable (Young, 1993). Consequently, and more importantly, communicative theorists are troubled by the impractical goals of the ideal speech situation and common good because they do not seem to undergird all policy discourses across the diverse interests of society (Young, 1997). While "the search for the common good in risk communication is an essential element in building public confidence in negotiation and mediation as alternatives to litigation contests that stress differences and diminish trust," it might be too democratically ambitious and therefore obscure and misrepresent all people and positions at risk (Chalk, 1987, 11).

An ecological rerationalization must be the product of a relationship between free consenting individuals and an inclusive community. According to Beck, by giving consideration to the multiple hazards of modern society and numerous dimensions of inappropriate science and technology, communicative action can be found in the modern age of living with unwanted risk. For Beck, all persons should be treated as ends and must not be managed by scientific experts. This is seen when Beck stresses correct reasoning for defining the good because traditional rationality alone has been unable to do so. Beck writes:

> Only by breaking the law of unseen side effects, by elevating decision making on technologies to public and political processes before and during the genesis of hazards, can we return the fate of the hazard civilization to the realm of action and decision making. We must reverse the prevailing practice of developing and financing new technologies first, and then investigating the effects and the hazards, and then finally discussing them under the guillotine of manufactured objective constraints. Only in that way could we minimize the hazards and open the possibilities for people to have a say in the matter. This time (for once) the political and democratic opening of hazard technocracy is also a way to prevent the hazards (Beck, 1997, 363).

Beck asserts that the traditional approaches to the study of the political condition are too narrow in scope and need richer and more real experiences to help examine society. For Beck these themes are: individualization, environmental discourse as technology, and the transformation and cultural renewal of the political (Beck, 1994).

Beck seems to discern a risk rerationalization for this ecological age. Rather than seeing risk communication as a transmission of information among members of Habermas's homogenous community, Beck views discourse about risk as common to everyone in the modern epoch. More importantly, this discourse can aid in communal reasoning for direct action. Beck believes in the need for regeneration in a society becoming idle, and therefore seeks to produce an autonomous knowledge that could cause an individual to reflect on his or her changed social condition and participate in environmental politics in a new way.

Beck's notion of a new "risk calculus" for citizen inquiry successfully rejects both technical elites who take their stand "for" scientific progress and public sceptics who are theoretically "against" it, and maintains that it is no longer possible to separate the two positions (Beck, 1997, 344). Beck suggests that each presupposes the other, and for this reason political interests influence scientific, technological, and societal development. Furthermore, Beck argues that because the environmental crisis is not a scientific crisis but a social one, it can raise society to a higher level of self-critique if more citizens were to realize that the communication of risk could only be done in a community. Beck explains:

> That means, however, that risks not only presume decisions, they ultimately free up decisions, individually but also in a fundamental sense. Risk issues cannot be converted into issues of order, because the latter suffocate, so to speak, from the immanent pluralism of risk issues and metamorphose surreptitiously behind the facades of statistics into moral issues, power issues and pure decisionism (Beck, 1994, 10).

Beck's moral construction is not like that of many post-modern liberals who leave the discussion of "the good" ambiguous, resulting in unlimited debate over philosophically limitless "rights talk." Instead, Beck creates a new democratic liberalism of communicative democracy. Unlike Habermas, for Beck there are no universal norms for modern society; rather there are many applicable virtues for humanity in the risk society. The "good," for Beck, is a neutral framework that is discovered when individuals reflect on the risk society, rather than just being rationally administered into living in the scientific safety of the modern society.

Perhaps the strongest explanation of communicative democracy comes from Iris Marion Young, who disapproves of organizing decisionmaking on communications that privileges the "best," or most convincing argument. In her opinion, articulate speakers tend to be dominant patriarchical leaders who are advocating traditional politics, mainstream economic status quo positions and inequitable societal relationships (I. M. Young, 1995). If discussion-based democratic models truly want to integrate the diverse values of the plurality of groups involved in risk deliberations, they must not marginalize non-traditional political actors. Within the area of risk studies, the most dominant speaking roles that could intimidate Habermas's implicit rules of deliberative democracy would be the economic elites using the language of statistics, economics, and scientific risk analysis. Therefore, in order to create an administration that is open, a policy design must integrate non-traditional political actors and stakeholders.

As a result of the dominant speaking style of technical experts, most risk communications eliminate public discourse by intimidating non-deliberative speakers. Ergo, a true Habermasian democracy would allow for an open discussion of policy topics in which unskilled orators could talk. Improving risk communication regarding agriculture is a concern of multiple individuals in the research and policy community in the United States and we would be well served by questioning:

> What's wrong with the agricultural research community? It's too damn scientific. It's not broad enough to consider social and ethical issues. Scientists are too far removed from the end use; they are ignorant or their own social context. Specialized language and concepts inhibit interdisciplinary research, to say nothing of the dialogue with non-specialists. Specialization is the only way we know how to do science, but agricultural science should include humanistic specialists as well (Stout & Thompson, 1991, 276).

In the brave new world of food biotechnology, policy decisions must not be exceptions to democratic practice. Ordinary citizens have a stake in the outcomes of such technical policymaking and have important views and insights to contribute to it (Laird, 1993).

An important question that did not seem to be communicated in the rBST case is, quite simply, what is to be gained and what is to be risked in the approval of a synthetic bovine growth hormone? For example, many dairy farmers did not look forward to rBST because studies suggested that up to half of them would be economically coerced out of farming because of falling prices due to the larger surplus of milk (Shulman, 1990, 111). Furthermore, the view was that rural communities would not benefit because the loss of large numbers of farmers would threaten their employment picture and tax bases (Shulman, 1990). Because real benefits would only be experienced by the few companies that are associated with large agricultural capital expenditures, urban areas would not realize any impacts of the contested dairy technology, either (Barnes & Nowak, 1990). Consumers also would not receive any economic reimbursements and for the most part would be wary because they have been taught since birth that milk is an unadulterated natural substance, the very symbol of nature, health, and growth (Barnes &

Nowak, 1990; Bultena & Lasley, 1990; Shulman, 1990). It is quite difficult to find arguments supporting rBST approval that seem to be deduced from the standards or the principles of communicative democracy.

Despite the overwhelming number of arguments pointing out the negative social consequences of the use of rBST, few, if any, policy experts argued against the role of technology in America's modern agriculture. Yet technology does not exist in a vacuum. It interacts with contemporary social, economic, public and personal realities (Barnes & Nowak, 1990). The acceptance of rBST was never openly debated in these multiple arenas. This is unfortunate, because research has concluded that issues "pertaining to hazards interact with psychological, social, institutional and cultural processes in ways that can heighten or attenuate individual and social perceptions of risk and shape risk behavior"(Renn, Burns, Kasperson, Kasperson, & Slovic, 1992, 137). How rBST came to be a rational administrative decision and communication for commercial authorization without debate over its broad effects is important to understand if for no other reason than that such an illustration could help us avoid future malpractices of risk communication.

Beck's theory of communicative democracy could jettison traditional political decisionmaking and establish a systematic alternative to rational choice theory in policymaking (Johnson, 1991). In critical social decisionmaking strategy, not only would science be evaluated, but the economic and social aspects would receive attention also. Recognizing the difficulty of importing the abstract philosophy of critical theory into public policymaking, many scholars suggest that critical theory already operates within the established structures and institutions of politics. Perhaps this is best noted in the preface to the Industry Risk Communication Manual that states: "Many companies are beginning to come to grips with what it means to involve citizens in environmental decisions that affect them"(Sandman, Hance, & Chess, 1990). This is essential because risk research has suggested for some time that values, attitudes, social influences and cultural identity are integral considerations in assessing, managing and communicating risk (Douglas & Wildavsky, 1982).

A new communicative framework of risk formulation can illuminate the organizational networking, scientific limiting, relationship building, and ritualised bargaining of analysts who work in policy and risk processes. The applications of critical social theory can be easily found in the multiple languages, discourses and frames of rhetorical claims in the public understanding of administrative and policy analytical practices (Fisher & Forester, 1993). A former Monsanto executive involved with the decision of rBST explained this best by stating:

> Agricultural biotechnology has mostly been a subject of debate. In a debate, someone is going to "win," and someone else is going to "lose," and the question of who wins and who loses may depend more on clever tactics and "debater's points" than on the

> underlying merits. Debates are characterized by attack and counterattack, by thrust and parry. Dialogue, however, is about questioning, listening, reflecting and understanding (Shapiro, 2000).

Critical theory is jumping out of the theoretical books of political philosophy and appearing as a guide for more meaningful political practice (R. Bernstein, 1995). As seen in the above analysis, the language of opposites and extremes characterized much of the debate and de facto decisionmaking in American public policy on agricultural biotechnology. Critical social theory, however, would teach policy stakeholders that with the specific issue of agricultural biotechnology and the general topic of risk, solutions are not either/or propositions. Therefore, in order to deal with this issue, reasonable decisions need to be made in the process of collective dialogue.

The written work of critical social theory is full of powerful psychological insights and extraordinary philosophical ideas. Unfortunately for environmental political theorists and policy scientists, these insights and ideas do not constitute a systematic theory of the good or just society, nor do they provide a model to the discipline of public policy. Nevertheless, despite explicit absence of a textual political treatise, many of the ideas of critical social theory can be constructed into an implicit political theory. If the idea of Habermas's open administration were truly considered by risk communicators, environmental democracy could become a reasonable liberalism. Through an analysis of the case study of rBST, administrators should be convinced that wider definitions of rationality, public health and safety are necessary. More importantly, implementing a discursive and communicative notion of risk communication may be the only way to consider the social benefits and risks of food biotechnology that can transcend the traditional science, politics and economics of today.

CHAPTER SEVEN

LIBERAL COMMUNICATION NEEDED FOR RATIONAL PRACTICES

1.CONCLUSION

The issue examined in this book has been agricultural biotechnology and the use of technical rationalism in the assessment, management, and communication of risk. It has been my recurrent argument that although the practice of risk analysis among industry technicians and governmental administrators is based on the ideology of liberalism, it is an impoverished or not fully performed form of liberalism. At stake in this poorly assessed, managed, and communicated risk analysis are billions of dollars in investments by the biotechnology industry and American farmers who have rapidly adopted varieties of corn, potatoes, cotton and soybeans that have been engineered to produce higher yields or resist pests. As a consequence of the American bureaucratic culture not learning any lessons from the dairy debate, the significant public policy divorce of rational-scientific risk and liberal-social risk continues today. The lack of public understanding, consent, trust, and discourse about the role of genetically altered milk in the future of agriculture is a prime example of the need for a broader analysis of risk that appreciates democratic values. Inclusive political analysis must become part of democratic life in America because the opinions associated with proponents and opponents of risk analysis are laden with ethical implications for the entire society.

The thesis of this chapter and conclusion of this book is that risk communication and risk analysis should be restructured as a reasonable political practice constituted by a theoretical combination of empirical science and normative philosophy. In the end, decisionmaking on environmental issues should only be possible if societal risks have been understood, evaluated, and prioritised in a democratic community. It is my suggestion that this can only occur if both the technical experts and the concerned public realize and appreciate the dissimilar political values each have with regard to risk. My hypothetical framework for examining risk analysis, in Chapter One of this book, argued that risk decisions about contested agricultural technologies are performances or arguments of liberal political theories. This was charted as follows:

Table 5. Risk Analysis

	Risk Assessment	*Risk Management*	*Risk Communication*
Rational Practice	Rational Choice	Administrative Decisionmaking	Institutional Participation
Liberal Theory	Utilitarianism	Social Contract	Critical Theory

As such, this book was to examine: (1) the methodology of rational choice as the practice of risk assessment and the performance of utilitarianism; (2) the theory of social contract as the argument practiced in the administrative decisionmaking of risk management and; (3) the institutional participation in risk communication as a performance of critical social theory.

I have shown that although risk analysis has been traditionally divided between a rational-scientific understanding and a social-liberal critique, they both originate from the same political philosophies of modern liberalism. Furthermore, even though they will always remain divided, it should not be overly problematical to get each side to succeed in respecting one another enough to have a reasonable discussion on the politics of risk. Whereas allies and adversaries of the process of risk analysis speak in different political languages, they have some core similarities in their political values. Thus, my concluding matrix for the ethical understanding of the issues involved in risk—seen in Table 6—is not all that dissimilar than the original one that I projected at the start of this research project:

Table 6. Risk Values Openly Debated

	Assessment of Equity	*Management of Power*	*Communication of/for Power*
Scientific Practice	Lowest Risk for the Greatest Number	Administrative Decisionmaking	Information Transfer
OPEN PUBLIC	SPACE FOR	REASONABLE	DEBATE
Social Theory	Greatest Good for the Greatest Number	Consent to Power or Trust in Power	Deliberative Participation

In assessment, the first stage of risk analysis, equity—having or dealing fairly and justly with all concerned parties—is the vital topic debated between scientific principles and social values in dairy milk production with genetically altered organisms. As explained, coming to a consensus on or determining a solution for arguing parties is multifaceted, because while the technical community measures risk by optimising the utility of the lowest risk for the greatest number, the general public views risk in quality, nature, or status and simply wants its milk to be good for the greatest number. Although the rational-scientific group and the liberal-social cohort both operate from a modern liberal utilitarian framework, they are not identical because the former views things with Bentham's mathematical value and the latter sees things with Mill's logical denotation.

In Chapter Three, I explained that Sandra Harding's criticism of the philosophy of science may allow both parties to begin to develop some consideration for each other's mental picture and perhaps commence altering their own dominant viewpoint or critical perspective. That is, technical advocates of Bentham might observe that their liberalism has an erroneous focus on objectivity and social followers of Mill may discover the elitism involved in their Tocquevillian position on political participation. Such mutual insights will destroy "either/or" thinking; that is, contending that environmental risks in modern agriculture are best defined either by experts or the public. Finally, this new paradigm of thinking should lead to the knowledge that the projected objectivity of the risk assessment is more like the expected subjectivity of risk management.

With respect to the bureaucratic operation of risk management and the evaluation of the uncertain consequences of rBST, power is the fundamental question. In modern liberalism, power—the capacity to get people to do something that they would not otherwise do—is established by the theory of social contract and the practice of consent or trust. The modern liberal John Rawls can be seen as the undergirding theorist for industry and governmental justifications of power in risk management. And the liberal-social opposition to the scientific-rational understanding of risk management is best explained by the critics of Rawls's political theory. Rather than endorsing a mathematical assessment of the highest expected utility, Rawls suggests a bargaining that allows decisionmakers to consider the most disadvantaged people in society. In order to make political institutions that favour the difference principle and use the maximin rule, Rawls endorses the power of the rational community. Despite his equalitarian intent, Rawls eliminates consent by creating an institution of elite decisionmakers who make social choices for the many within an original position of strict rationality and a veil of ignorance that focuses on the cost-benefit goals of society.

Similar to Rawlsian apologists, risk managers argue that a well-ordered and safe polity will be chosen in a just and fair manner under the scientific rationalism of formal modelling. Critics of the process of Rawlsian difference politics and risk management, however, claim that it is unreasonable to organize society by the subjective principles and probabilities of asocial experts who often exhibit an overconfidence bias that maintains that there will be no serious problems or negative health effects from a new justice or technology constructed only with apolitical rationalism (Harsanyi, 1975; Pojman, 1997). In Chapter Four, I argue that social critics of risk who advocate regulatory reform and participatory policymaking in order to combat the power of risk managers should instead develop long-term political solutions by creating a more liberal society with classical ideas of private consent of the governed and public administrations of political trust.

In order to achieve some level of political trust in risk management by the reasonable public regarding contested agricultural technologies, I build on John Locke's notion of individual tacit consent and Thomas Hobbes's idea of public sanction. Additionally, some insightful strategies for respecting the knowledge of the other's ideas of environmental policymaking can be discerned from feminist political philosophy. According to some feminists, the outstanding characteristic of democratic theorists is the hopelessly unrealistic—albeit classical—belief that citizens participate in politics when political elites compete for their consent in stable institutional arrangements (Pateman, 1970, 3, 20). The bulwark of democracy for Pateman is not contractual representative institutions but a participatory society, because while all social contracts result in the loss of civil rights, liberties, and powers, civic participation or citizen knowledge develops, fosters, and establishes trust. When critics of the technical judgment of risk realize that their liberal intent at a social contract is an impoverished or distorted form of liberalism rather than an authoritative power deployment, a meaningful discussion of reform and trust may emerge. For Pateman, this new democratic society is characterized by "the central assertion that individuals and their institutions cannot be considered in isolation from one another" because true democratic liberalism must be at "the level of socialization" taking place in all spheres of public and private life (Pateman, 1970, 42). For a modern liberal polity to completely function, it is necessary that democratised citizens and experts understand that all social systems are political systems with the potential for fuller participation. This is best illustrated by the inclusive communication of the many individuals involved in contemporary risk politics.

In risk communication, it is important to distinguish rhetoric—dissemination of rational information about a scientifically determined risk decision—from democratic process—the civil determination of a social risk.

As discussed in the previous chapter, in the final stage of risk analysis, the central issue in dispute between scientific principles and social values is whether risk communication is to be a one-way or two-way conversation. In the risks associated with food biotechnology, the argument is between communication as explanation of an already decided risk decision and communication as deliberation of an undecided risk decision. As one might expect, it is not easy to find a middle position in the problematic relationship between the institutional obligation of transferring information to the public and the democratic insistence that all social decisions of risk be subject to deliberative discussion. Fortunately, however, that relationship is easily understood because both are rooted in Habermas's theory of communicative action. Although the rational-scientific group and the liberal-social cohort both operate from a social critical theory framework, they are not identical because the former stresses Habermas's pragmatic value and the latter emphasizes Kant's ethics.

Habermas's commitment to developing a universal perspective on rationality and ethics is what is noted by the rational elites in advanced capitalism and administered democracy. In their interpretation, Habermas suggests people should work for rational competence and communal knowledge should privilege the linkage of fact and value. According to their scientific worldview, Habermas's theory of communicative competence or "universal pragmatics," the ideal speech situation is a "practical hypothesis" or normative structure for rational discourse because it defines the conditions for political arrangements that unite practical and theoretical knowledge claims. That is, the ideal speech situation is a pragmatic tool for guiding decisionmaking on what the rules for information transfers should be like or how policymaking about public records or rules should be determined. Rational-scientific readers of Habermas's critical social theory do not accept the idea that Habermas's discourse ethics is to be a prescription for a radical new politics of consensus on the part of an inclusive community.

According to advocates with a liberal-social perspective of risk, however, Habermas's theory of communicative action has progressive implications for a new theory of ethics and a new practice of politics because they want to establish Kant's ambitious idea of a free community without a universally defined common good. Jurgen Habermas learned from Immanuel Kant the idea of the dignity of the individual and the concept of community without the loss of freedom of being determined by a defined common good. Kant maintained that there are no universal laws of right and wrong; instead there is only the "categorical imperative," an internal moral law regarding how a person ought to act for him or herself and within a community (Osborne, 1992). Every free moral action, therefore, is in accordance with the common good, which enables each autonomous person to exist together with the freedom of all others according to the universal law (Curtis, 1961b). For social-liberal advocates of risk analysis, Habermas is the key to creating a risk society that is free and ordered because he respects all risk positions in his speech ethics.

The ideal speech situation is a condition of discourse free from domination, with equality of participants for the objective of reciprocal understanding. Politically and socially, this leads to Kantian notions of equality before the law, the need for correct legal procedures, and social norms that would increase political involvement. Habermas, like Kant, optimistically believes the human condition shows a neutral design for a perfect modern liberalism. Thus, the ideal speech situation is meant to establish a new society, politics, and ethics with a comprehensive theory of modern liberalism achieved only by consensus through undistorted and unconstrained discourse between all members of a community.

After reading Young's criticism, however, both parties can acknowledge each other's interpretation of Habermas—rational followers of the conservative Habermas will note the elitism involved in their status-quo liberalism and radical readers of Habermas will observe that their liberalism has a utopian emphasis on consensus. Such insights have allowed me to find other critical social theory thinkers who will aid in developing a better understanding of future risk communications. I have argued that Marcuse offers a significant amendment to the understanding of deliberative democracy and Beck provides a noteworthy addition to the thinking behind communicative democracy. These are two important concepts in creating a theoretical case for a reasonable environmental liberalism. This is important because even though democracy may be established by a foundational philosophy, it will only be sustained in the field of environmental policymaking with an active citizenry and a liberal bureaucracy (Barber, 1984, 1994; Ferre & Hartel, 1994).

The concluding analysis of this book is that the foundational liberal theories of risk should be better understood in order to rerationalize the practice of risk. However, it should be noted that the liberal and the rational interpretations of risk should not be completely collapsed into one another because they perform different, yet necessary, functions. Rather, in order to combat the current undemocratic character of environmental policy, each group should understand and consider the opposing group's position. If scientific practitioners of risk clarify the values of their bureaucratic agencies, American society might be able to correctly interpret and appropriately react to the risks it actually incurs with respect to food biotechnology. Additionally, defining the liberal objectives of the American public will allow bureaucracies to appreciate the reasonable concerns of society and, consequently, compel the public to come to terms with the limits of administrative practices.

At the most rudimentary level of understanding, this book has explained two commonly opposed views of environmental risk. The first view maintains that sound judgments on modern agricultural policy are established in accordance with canons of good scientific rationalism and the second asserts that good decisions about biotechnological farming are to be established by valid social reasoning (Lehman, 1995, 107). In this book, I would have preferred to be a policy facilitator rather than a theoretical analyst. For example, I would like to tell the scientific-rational community that a prerequisite for rigorous scientific research and rational utilization of advanced biotechnological methods of agriculture is a thorough analysis of all of the ethical, social, and economic consequences (Lehman, 1995).[40] I would also suggest that a "GMO ingredient free" governmental label is a moral imperative for establishing consumer consent and citizen trust in the brave new world of genetic foods. Finally, I would insist that governmental agencies replicate industry risk assessments and communicate results, not management decisions, to the public for an open forum of deliberation about societal risks.

Additionally, I would like to inform the liberal-social community that discussion of "risk free" or "natural food" should be eliminated because they are agriculturally impractical. I would also declare that all risk decisions could not be made with consensus because environmental democracy is too ambitious and cost-benefit analysis is not incorrect just incomplete. Finally, I would suggest that biopatents are a legitimate intellectual property right and the populist accusation of corporate domination of farming in America is not helpful in communicating environmental risk issues. More important than offering short-term policy directives, this book has attempted to suggest a philosophically informed long-term procedure for risk analysis in modern farming.

This philosophical process will help to create a middle political space for social and ethical debate. This middle position between rational practitioners and social thinkers will allow decisions on individual cases of environmental risk to be made both with universal standards of technical definitions and limited public perceptions. It does so, however, within the context of modern liberal frameworks that are attempting to rerationalize scientific experts who speak of risk as the probability of harm associated with a given product and practice, and offer reasonableness to environmental social-liberal publics that understand risk as the possibility of loss, injury, or peril to the well

[40] Please also see Lehman's "Concerns about the Ethics of Genetic Modification" presented at the Fifth World Congress on Genetics Applied to Livestock Production, referenced in Lehman (1995).

being of one's general situation (P. B. Thompson, 1995). As a consequence, however, of using scientific knowledge for interpretation and viewing social knowledge with flexibility, I am left with a viewpoint that is by no means absolute.

My conclusion that modern liberalism is the "solid, systematic standpoint or core of moral concerns from which to criticize or oppose any particular patterns of technical development" does not answer all risk questions (Winner, 1993, 374). Although not a universal solution, such an idea of reasonable liberalism does, however, suggest that "choices are available; that the course of technological development is not foreordained by outside forces but is, instead, a product of complex social interactions"(Winner 1993, 375). How specific risk decisions are going to be made with the rerationalized principles of modern liberalism is open to discussion and beyond the parameters of any individual study. Thus, even though my ethical matrix has not offered closure to all of the debates surrounding the risks of food biotechnology, or even rBST, it has hopefully offered a new course.

2.CONTRIBUTIONS OF THE STUDY

Analysing, measuring, or discussing to what degree the United States is democratic is commonplace in American political science. In fact, there is an embarrassing amount of noteworthy scholarship dedicated to examining the democratic nature of American culture, politics, and society. The peaks of the landscape of academic political science—from de Tocqueville through Dahl, and Lowi, to Mansbridge or Putnam—are theoretical interpretations of the "democratic" or "undemocratic" character of the United States political government. Furthermore, my assertion that environmental risk practices are the product of a rational-scientific theory of liberalism is also commonplace because it has been noted by many authors that environmental politics, particularly, and American politics, generally, function as a form of democratic elitism (Lowi, 1979; Mills, 1956; Ophuls & Boyan, 1992; Paehlke, 1994; Parenti, 1995; VonSchomberg, 1993; Wildavsky, 1989). The United States, for instance, simultaneously has an emancipated, but uninterested, citizenry and can be characterized as a free polity with a limited power structure. Although the United States has established democratic institutions with inclusive procedures for governmental activity and has American citizens who express pride in their nation and tend to be very opinionated and vocal about politics, the actual level of active and knowledgeable political participation by non-political elites in post-industrial America is insignificant. That is, even though the potential for self-government and the presence of aristocracy are within American democracy, they often are not held in equilibrium.

The American system of decentralized and fragmented pluralism offers the promise of self-government because of its openness to multiple sources of political mobilization, but the underlying reality is that influence is organizationally concentrated in the hands of a few policy stakeholders (Pateman, 1970). In surveying the political literature concerning variation in the extent of American democracy, the first observation is that United States public policy can function without a lot of citizen interest, knowledge, and participation (Nagel, 1975). The quality of such American public policy and its corresponding "democracy of the few," however, must be questioned. The second observation is that while the lack of participation in American democracy has structurally changed throughout history, it has been an enduring condition. For instance, although elites have always managed democracy in America, the identification of the elites has changed. That is, originally a social aristocracy managed the American enterprise; then, during industrial capitalism, it was an economic aristocracy; and, now it is, to some extent, an aristocracy of knowledge (Enteman, 1993). More importantly, even though the ruling elite has changed its undemocratic influence on the American polity over the years, liberalism has always been the foundational political theory of American government and society. Whereas the dominant ideology in American environmental policymaking today might be described as some elaboration on democratic elite theory, its political process is identical to all politics in the United States—modern liberalism. As such, this book, although about democracy, has attempted to look beyond simple conversations of participation modelling in public policy and toward the complex tradition of the philosophy of liberalism that undergirds American public policy.

If the crude flirtations of status and money once tempted policymakers to stray from devotion to majority consent, today their heads are turned by the infatuation with the charms of specialized information and technical knowledge. The democratic theorist Sartori once remarked that contemporary American public policy seems to have been "uneasily, and often hastily placed between two extremes: a government of non-experts over experts, or a government planned out by experts without democracy"(Sartori, 1987, 431). The tension between the ideal democratic theory of modern liberalism and the elite practice of that same modern liberalism is a major problem with democratic government today, says Sartori, because technology increases the gap between the non-specialist citizen and the powerful expert in government (Sartori, 1987). Quite simply, our highly technological society makes us ever more dependent on the judgments of experts. What, after all, does the ordinary citizen know about the causes of or solutions to environmental degradation, the use or misuse of toxic pesticides, the regulation or deregulation of farm subsidies, the proper agricultural trade balance between the United States imports and exports, or whether public policy should permit genetic manipulation in modern food production?

Democracy and technology have not always had an easy coexistence. Recall, for instance, in the less technologically complex times of Plato, Socrates explained that a polity was like a ship afloat perilous waters and therefore must be "commanded by the most competent pilots; to allow the crew, ignorant of the art of navigation, to participate in running the vessel would be to invite shipwreck" (Ophuls & Boyan, 1992, 210). At first glance, Plato's conclusion appears to be obvious. When the involved parties—citizens or crewmembers—differ greatly in competence, the safety of the ship or the society requires that the democratic value of involving everyone be subordinated to the benefits obtained from competence on the part of the skilled elite. However, as I have previously argued in this book, Plato's analogy and corresponding advice is not totally applicable, because even though America might be on a dangerous sea voyage, the few pilots making decisions for the many shipmates are not competent philosopher-rulers operating with complete certainty of their art and craft on behalf of the common good of their craft or country. Additionally, while all societies have been divided into the "few and the many," all of the citizens of the American polity, unlike most others, operate on the same political ideas of liberalism and, therefore, do not have to function separately. The powerful few are the rational-scientific experts in government and industry and the many are the democratic citizens who are uninterested in modern public policymaking, yet both think about policy problems and solutions within the structures of modern political liberalism.

It has not been overly important to differentiate the theoretical nuance of modern liberalism from that of classical liberalism because both the rational-scientific practice and the liberal-social theory of risk are generally concerned with modern liberalism. Early in this book, it might have seemed that biotech companies were functioning from a locus of classical liberalism, but actually industry-government risk practices were based on the ideas of modern liberal thinkers: Bentham in risk assessment, Rawls in risk management, and Habermas in risk communication. Although the public, environmental activists, and I might all like to see these modern liberal ideals fully realized or even augmented with different philosophical thinkers within the same tradition, both the current practice and desired theory of risk analysis are functioning from the same foundation. More important, because they have same central belief system, they can easily respect their differences in risk analysis, and given particular contexts, implement the opposing group's theory and practice.

Notwithstanding public opinion, the core of most debates about contested agriculture and risk policy in modern farming is not found in any difference between classical liberalism and modern liberalism, but rather is concerned with modern liberalism realizing its politics of scarcity or ecological limits (Ophuls & Boyan, 1992). More important, some of the environmental problems associated with modern liberalism can be discerned by investigating the political theory foundations of risk analysis. That is not to say, however, that the development of modern liberalism from classical liberalism does not offer some lessons for today. Recall that at about the turn of the twentieth century, the demands upon classical liberal government—which stresses an individualistic perspective of natural rights allowing for only a minimal level of governmental involvement in economic affairs—increased, and many intrepid thinkers argued that liberals must assume a more activist attitude towards social issues (Curtis, 1961b). Just as when classical liberalism noted that its formal commitment to natural rights had the substantive effect of protecting distinct social advantages, modern liberals today must note the ecological limits of their neutral and utilitarian frameworks. Therefore it has been my subtle suggestion in this book that the philosophy of modern liberalism can make such an ecological discovery by fully performing the theories of risk analysis.

In reflecting on the dual influences of modern-democratic-liberalism in the values surrounding risk analysis, one questions whether the dominant philosophical nature of the rational-scientific view—embodying primacy of technical knowledge as seen through private science and limited government—or the reasonable-social understanding—embracing the primacy of the common good as seen through public consent and governmental trust—will be used in making risk decisions. That query is possibly the genesis for all past, present, and future political debates of American environmentalism, I have argued that such thinking is problematic because it assumes that each of the two political factions can operate alone. That is, believing that risk decisionmaking either should function with more citizen participation or with greater technical understanding is the product of thinking within a false dichotomy. Using science or society for understanding risk is not an "either/or" proposition, because they necessarily operate together. Although they have operated in contradiction to each other in the evaluating, marketing, and regulating of rBST, that does not have to be the case, nor should it be.

This book has concluded that science and society should, however, function in unison and not in conflict. I do not believe that science and society can or will be easily wed, but I assert that science must not conduct itself in a lab divorced from its social context and, society must not dictate behaviour separated from the knowledge of the actual dangers and benefits offered by science. Those two competing political ideologies are both the foundation of undemocratic environmental policy and the solution to rerationalizing or liberalizing American risk policy. For instance, public policy programs that are considering using contested biotechnologies should incorporate some of the ideas of the democratic polity before making a scientific decision. Although such a modern liberal convention may alter the predetermined ideas of the technical community, experts should realize that in eliminating some of the authority of rational-scientific decisionmaking they would be building trust in liberal-social theories of American policies.

American bureaucracy, therefore, should work with citizens rather than for them. If policy sciences were to realize that their disciplines could be the perfect blend of theory and practice, modern liberalism could be realized and bureaucratic and technical experts would no longer just have to manage the reasonable voices of the American polity (R. E. Goodin, 1982). That is, when empirical risk science and ethical risk theories are used in tandem, they guide public policymaking and make modern liberal democracy work.

The academic benchmark for much of prevailing American political science discussions on the "democratic" and "undemocratic" nature of the United States has been *Democracy In America* by Alexis de Tocqueville. While de Tocqueville offers many noteworthy descriptions and prescriptions, the most important to this conversation is that because America operates on the single principle of liberalism, it lacks a genuine revolutionary, reactionary, and real political culture. According to a popular interpretation by Louis Hartz, best seen in *The Liberal Tradition in America*, because Americans "have arrived at a state of democracy without having to endure a democratic revolution; and that they are born equal, instead of becoming so," they have developed a political arrangement that is indifferent to challenges of feudalism, socialism, and communism (Hartz, 1955, 35). Hartz's principal deduction from de Tocqueville's work—that there is a political homogeneity of liberalism caused by a non-radical Constitutional birth, leaving contemporary America with an ideological rigidity of conflict and puerile politics—deserves some attention. Clearly, there are a myriad of "liberal" principles and structural devices to protect limited government, property, and minority rights, contrived by both the framers of the Constitution and later political decisions, but it would be erroneous to overlook the depth and scope of political ideas within the single ideology of American liberalism (McDonald, 1985).[41]

Consider, for instance, *The Federalist,* which is arguably the most authoritative commentary and account of the hopes and fears of the authors and supporters of the Constitution of the United States. Although the intention of The *Federalist* papers was to explain and defend the Constitution of 1787, its recurring, if not constant purpose is, in James Madison's words, to find "remedies" for the "diseases most incident to Republican government." In *The*

[41] For a general discussion of the philosophical traditions involved in making American liberalism, please see McDonald (1985). Specifically, note the debates surrounding the contention that liberalism is the only political theory involved in the framing of the United States Constitution.

Federalist No. 10, perhaps the most famous of the 85 essays, Madison takes up the argument that anarchy, war, and majority rule will plague a free government or a pure democracy. Madison concludes that a dominant faction or a tyrannical majority are the benefit and risk of liberty and must be controlled rather than suppressed. In the course of a long exploration, Madison argues that a divided, balanced, and limited government with a system of representation can control a republican majority. Although a sophisticated classical liberal thinker, Madison was a politician, not a political theorist (Ceaser, 1990). As such, he compromised on ideas of classical liberalism and created an unprecedented ambulatory document that was able to recognize many modern liberal ideas of human freedoms, rights, and equalities. Furthermore, there is no reason to believe that such liberalism could not expand to accept the limits of our current environmental practices because there is just as much or as little of a rigid ideology in America today as there was at its conception.

Although de Tocqueville is correct when he asserts that American politics is narrow, Hartz is wrong when he contends that all we have are narrow politics. The American political arena consists of two discussed ideological versions of modern liberalism, leaving an ambiguous middle that cannot hold its own and is the residence for everything between libertarianism and populism. A matchless illustration of the diversity of American politics is the interminable list of labels used to simply define the American structure of government: democracy, representative democracy, constitutional democracy, liberal democracy, republic, representative republic, and constitutional republic (deLeon, 1995). My intention in discussing the political theory of *Democracy in America* at length is not to assert that, as a result of his thesis being commonly misinterpreted, mine is to be given equal consideration. Nor am I asserting that this book is mainly talking to de Tocqueville or his followers. Rather, I am contending that like the classical liberalism that de Tocqueville critiqued, the modern liberalism found in the rational-scientific design of risk analysis today lacks any cohesion and universality and, thus, can be developed into a different form of American liberalism.

Paradigm shifts in the study of politics and the policy sciences are periodically required. The field of political science should terminate inquiry into what political science could do in society and begin asking what it should do (Ceaser, 1990). American political science, in some sense, must be engaged in the maintenance and, therefore, criticism of the political regime within which it exists. Louis Hartz's analysis—there is an American political homogeneity of liberalism which results in an ideological rigidity of conflict and puerile politics—is more limited than flawed, but seems to explain why American individuals feel communally insignificant, despise political institutions, are bored with political parties, and fail to engage in technically complex public policy debates. More importantly, the idea that American liberalism is limited might explain why the scholarship of political science has become normatively vague, discussing objective realities and statistics. Journalist E.J. Dionne, Jr.'s book, *Why Americans Hate Politics,* asserted that modern liberalism is incorrectly framing political issues in a series of false choices and thus preventing the nation from getting to the questions that truly plague them.

Dionne asserts that there has been a decline of a "politics of remedy" and group solidarity. Dionne concludes that political parties ought to do coalition building around the neglected centre issues and develop a social reform that commands support among more Americans. Contemporary risk theorists need to examine the purpose and course of the discipline of politics and should give Dionne some consideration and think about how to address the equivocal middle of the political spectrum, not in the electoral sense, but in the arena of political education. Political science scholarship should be more theoretical and begin to revalidate and strengthen the idea that competing factions clarify the alternatives and promote national cohesion and understanding. Additionally, some reflection should be given to *The Federalist* No. 9, which proposes a "science of politics" that would guard the new union from the disorders that had plagued Republican regimens of the past because, just as it did two centuries ago, political science today considers "free government was/[is] inconsistent with the order of society." As such, understanding the tension within American liberalism is one of the main objectives of this study. This book, at its best, has described recent decisionmaking in risk analysis with food biotechnology as being dominated by an impoverished modern liberalism and the rational-scientific community, and it has also attempted to suggest that a paradigm shift toward modern liberalism and more democratic values is needed in future risk deliberations about food biotechnology. I hope that, at worst, it represents a type of political science that reaches beyond daily examinations of politics and offers to the discipline a theoretical contribution about policy deliberation and implementation.

3.RECOMMENDATIONS TO POLICY SCHOLARS AND PLANNERS

A paradigm shift is needed in the contemporary study of public policy. With its atheoretical emphasis of positivism, policy science has become a political analysis without political theory and, for most scholars, only the study of particular programs. Public policy, once the concern of all of political science, has dwindled into a marginalized quantitative discourse of little interest to planners, citizens, and even other political scientists. There are a few policy scholars who correctly argue for a new focus—for instance, Torgerson argues for hermeneutics, Fisher for less technical knowledge, and deLeon for a democratic ethos—and they all assert that the future course and purpose of public policy would be best served by a more inclusive, postpositive, and interdisciplinary study. For me, however, a new paradigm would simply attempt to reunite theory and practice by orienting political philosophy as a guide for policy science tasks. This is best seen in a reasonable risk analysis that strives for order, but considers changes in scientific knowledge and social understanding before making an environmental policy.

My democratic position on risk itself cannot provide a blueprint for specific decisionmaking structures, but it can provide a standpoint from which existing procedures and structures can be judged. This research fosters the notion that if risk analysis studies itself, it will recognize that its dominant values, methods, and techniques have been created, developed, and nurtured within a closed political community. In mapping the correlations of public policy and political theory, I seek a self-reflective social science, or a risk analysis that looks for simplicity and then distrusts it. If public policy was to flourish along the set of dimensions offered by political theory, it could expand as it never has in the past. When scholars and citizens alike start seeing public policy as a topic that pervades nearly every aspect of politics and that should be interwoven into the entire understanding of the human condition, its benefits would prosper. In this spirit of political theory-public policy, I have two recommendations: one is to the university study of risk policy and the second is to the administrative action of risk policy.

Public policy—as studied and endorsed by academics—has been disturbingly unclear as to what type of democracy it favours. "Is it an indirect, Madisonian democracy, one of representation and factions, or is it a more participatory, deTocqueveillean democracy dependent on general civic participation" (deLeon, 1995, 886)? The historical and ongoing debate about the respective democratic visions of the American polity should probably never end, but good public policy scholarship should discern what would best repair the liberal deficit in American government (Dahl, 1956, 1983, 1989). Not surprisingly, the majority of the work policy science does today and has done through its history uses a Madison-inspired rational model of analysis that focuses on pluralism. Traditional liberal rationalism, as discussed in this study, talks of egalitarian policies and redistribution programs. Under this type of scholarship, the idea of politics of expertise is subtly endorsed by academics because the policy sciences function best as knowledge for a government that is separated from its subjects (deLeon, 1992).

Alternatively, what I recommend for the policy sciences is a focus on a participatory democracy, both normatively and empirically. The building of individual and civic virtues to remedy democracy might be first attributed to Jean Jacques Rousseau, for he is the author most referenced by policy scholars who desire a social-liberal emphasis in the policy sciences (deLeon, 1994). He is absent as a primary theorist in this study because Rousseau's ideas are deeply rooted in many of the secondary authors who endorse an environmental democracy for American politics (Cutter, 1993; Mansbridge, 1980; Ophuls & Boyan, 1992; Pateman, 1970, 1985; Plumwood, 1996; Stone, 1997). Furthermore, Harold Laswell—the first modern public policy theorist—called for a "Rousseau like" participatory emphasis in the policy sciences. It has been said that the father of the now dismal policy sciences "envisioned a profession which would promote both the general enlightenment of the population and the widespread sharing of power in democratic policy process. What we witness in the development of Laswell's work, then, is a methodological-political convergence which centers on participation …"(Torgerson, 1986). Laswell, like many early political scientists, believed the pedagogical preference for research questions that examine democratic participation could help diminish the democratic deficit among American citizens. Also, while this deficit is seen in many areas, "no wider gap exists today between lay citizens and experts in the United States government than the profoundly different and often polarizing perspectives they exhibit on what constitute the most severe threats facing public health, safety, and the environment"(Durant, 1995, 30). Although the effect of risk research on understanding or bridging this gap is not known, it is clear that if the entire research community is focused on studying the same type of democracy, the effect can be better than if the focus is diffused.

In the middle of her ethical valuations of environmental risk, Kristin Shrader-Frechette offers profound recommendations to political administrations dealing with risk policy. For instance, Shrader-Frechette once called for technology tribunals or citizen juries to inform elite policymakers about the concerns the public might have with proposed industrial processes or new technological products (Thiele, 2000). In a similar—albeit less ambitious attempt—to institutionalise public input into environmental risk, I would like to see the discussion of a single federal regulatory agency for food safety. Quite simply, developing an inclusive liberal risk assessment, management, and communication process for food biotechnology would be much more possible with a single regulatory body. Currently, the FDA, USDA, EPA and to a limited extent the Federal Trade Commission (FTC) oversee all aspects of food processes in the United States. Although each agency appears to have discrete and specific functions, the complexity of biotechnology has created a complex labyrinth of rules, regulations, and responsibilities that make rerationalizing risk analysis difficult. A new agency could take all of the disjointed food supervisory responsibilities of these agencies and oversee the marketing, labelling, and production processes of new and conventional foods.

"As commercial biotechnology emerged over the last few decades, its unique issues were merely absorbed into the current regulatory structure for food, drugs, environmental issues and agricultural production" (Beaudoin, 1999, 263). Consequently, there have been weird loopholes, incomplete definitions, and awkward jurisdictional battles between agencies dealing with modern food production. In the USDA's Final Policy Statement for Research and Regulation of Biotechnology Processes and Products, for example, the largest number of complaints involved: (1) the competing regulation of agricultural organisms between the EPA and USDA, (2) delays in risk management decisions about biotechnology drugs used in agriculture because of jurisdictional battles between the FDA and the USDA, and (3) the states acting independently because of no federal oversight over a myriad of new issues relating to food biotechnology that were not addressed within the previous framework (Beaudoin, 1999). Furthermore, in contrast to the current and recent administrations, a new food safety agency would not work with companies to bring products to market; instead, it would ease the application by making companies deal only with one agency. The creation of a single U.S. food safety administration—an idea not exclusive to myself—would increase public input because elimination of unnecessary bureaucratic overlaps and delays in the rational-scientific risk assessment, management, and communication process will make it more citizen friendly.

4.FUTURE RESEARCH IN RISK STUDIES

A notable omission in this study is any sustained attention to the field of comparative politics, both as way for understanding the natural resource policies of political administrations and as a methodology for discerning variations in the ecological values of political bodies in the course of their decisionmaking. This omission is important given the scope of food biotechnology issues that are relevant to international relations and comparative politics. Generally, it would be valuable to research, explore and revisit the influence of culture on political decisionmaking in risk analysis. Specifically, I think some research should examine why the United States is willing to accept new technologies, support relatively risky businesses, and establish flexible regulatory rules for the nascent science of food biotechnology, while European Union (E.U.) member-states have a sceptical approach to new technologies, a bias toward traditional business, and respect for moderate laws of regulation (Echols, 1998). Such a comparative study would not only offer a cultural guide for understanding the historical differentiation in agricultural policy between America and Europe, but also provide a theoretical framework that might explain why the United States and E.U. regulations are becoming more similar (CEC, 2000).[42]

Despite the globalisation of the food supply and the amount of international commerce involved in modern agriculture, cultural attitudes greatly affect the regulation and commerce of food production. Understanding the differentiating political acceptance of GMOs between the U.S. and the member-states of the E.U. is problematic because the decisionmaking apparatus for considering new technologies in both political contexts is the scientifically based assessment and rationally administered management of risk analysis. Although cost-benefit assessments that use scientific data to project probabilistic consequences of proposals and their alternatives have been found to have cultural influences, few researchers have done case studies in which culture, and not science, determined the final outcome of an environmental risk policy. It appears, however, that the reluctance to grow and import GMO food in the E.U. and the political willingness to grow, consume, and export GMO agriculture in the U.S. is largely explained by culture because the scientific conclusions about human health safety and environmental integrity of each ruling body are similar.

The controlling objective of good social science is to explain and predict relationships between social phenomena. For some time political scientists have offered evidence of cultural attitudes toward food biotechnology as the causal explanation for differing policies of food biotechnology between the U.S. and the E.U. While this data can allow other empirical comparative researchers to test new deductive hypotheses, they can also help generate a new interpretation of a political theory than can develop inductive concepts which can be utilized across time and space. Valuable political science research would, for example, explain why the E.U. is starting—however slowly—to follow the U.S. cultural model of risk analysis in terms of allowing many economic and technocratic considerations to affect environmental policy about food biotechnology. That is, despite a great deal of public deliberation against genetic engineering in modern agriculture in the member-states of the E.U., it appears that central decisionmaking by rational experts of the E.U. are starting to consider the regulatory regime of GMO food with economic ideas.

As a consequence of the practice of risk analysis not being completely explained by utilitarian, social contract theory, and critical social philosophy,

[42] See the CEC (2000) Final 719 COM Draft dated 12 January.

this book does not generate a conception of risk that adequately reflects all of the contemporary public debates over contested technology. I maintain that liberalism is the basic system that underlies risk debates about food biotechnology in America, but I question if my research design—which is concentrated only on different types of modern liberalism—has overlooked other important schools of environmental ethics and philosophy. This criticism is important because while attempting to write a book that may forge agreement on basic policy imperatives in environmental politics, political theory examinations of liberalism have been accused of being needlessly intolerant of the basic interests of political science.

In surveying all of the substantive knowledge in contemporary political science, the disciplinary professional John Gunnell has characterized the relationship between political theory and political science as a disconsolate one (Gunnell, 1993, 195). In his eyes, political theory has largely retreated from critique or even involvement with political science and, as a consequence, political science has afforded political theory the ultimate disdain of pure tolerance (Gunnell, 1993, 191). This situation has allowed many important issues of politics to slide into obscurity or to become detached from the practical concerns that originally generated them. Within the field of environmental ethics this can be observed by philosophers who have the capacity for intangible theorizing that transcends most human perspectives or experiences of ecological issues. The alienation of political theory and political science is, according to Gunnell, a result of the professional enterprise of political theorists. Under the guise of the intellectual principle, political theory pushed itself in the direction of its own discipline by encouraging internal conversations about the political philosophy of liberalism (Gunnell, 1993).

After its emigration to its own intellectual and professional archipelago, political theory makes only occasional visits to political science and largely only as a location for discussing its own internal discourse about liberalism and its inability to defend and maintain itself (Gunnell, 1993). Even a careless observer of political philosophy would notice that political theory does seem to be saturated with complex discussions about liberalism, even when the concept is not a focus of the original research question. I am interested in conversations between public policy and political theory because the former often has no critical dimension and the latter occasionally forgets that it can reflect on authentic political actions. Despite the undeniable significance of liberalism in political theory, I question whether my research agenda of attempting only to examine the liberal foundation of risk analysis was a product of what Gunnell calls political theorists' fascination with liberalism, or my bias of wanting public policy to work for democracy.

That being said, further research should question if political philosophies other than liberalism can be found useful in risk analysis. As one of the reviewers of this research has pointed out, although the relationship of liberalism to risk analysis is important, to focus on it exclusively risks overlooking the relationship between pragmatism and risk analysis. Pragmatism is marked by the doctrines that the meanings of conceptions are sought in their practical bearings, that the functions of thought are to guide action, and that truths are to be tested by the practical consequences of belief (Osborne, 1992, 140). A pragmatist's approach to problems, affairs, and truth seems like an obvious fit for risk analysis, but I am really interested in the association of pragmatism, science, and democracy. Although the American philosophical movement of pragmatism started with C.S. Peirce and William James, future research should really focus on John Dewey. For it was John Dewey who turned pragmatism into an ethical and political theory that many historians believed influenced American social policy, most notably in the fields of educational and legal reform.

Dewey is an important author to be investigated for American public policy because while he is a democrat, he does not reach that political stance through the dominant ideology of liberalism, but rather through the intellectual curiosity of science. Although his theoretical journey was an extensive and convoluted one, Dewey expressed his concluding philosophy in one sentence: "A general application of the methods of science, to every possible field of inquiry, is the only adequate means of solving the problems of an industrial democracy" (Fisch, 1951, 328). Seen in his larger work *The Public and Its Problems* and smaller pieces like "Science and Society" and "Creative Democracy: The Task Before Us," Dewey's philosophy is not a comprehensive theoretical system of thought, but rather a method of thinking. Such an alternative to traditionally liberal philosophies of environmentalism may aid the explanation, solution, and discussions of natural resource and ecological issues (Light & Katz, 1996).

The ultimate goal of the social sciences is to produce an accumulating body of reliable knowledge that can enable, explain, predict, and understand phenomena that interest humankind. Such a reliable body of knowledge could be used to improve the human condition. If political science is to be pursued both for its own sake and for improving human conditions, it must have explanations that admit their possible limitations (Wildavsky, 1993). Good political theory must therefore consider multiple cultures, beliefs, perceptions, experience, and observations in order to inform political science and the human condition. With that conception of social science, the limitations of the correlations between liberalism and risk analysis must be investigated. To that end, scholarship that examines the importance of Dewey's pragmatism and risk analysis should be completed.

5.SUMMARY

As an undergraduate student in philosophy, I was taught that the discipline was bifurcated into ethics and politics in attempting to resolve the question: "What is the best way for human beings to live, both as autonomous individuals and as agents within a community?" Ethics or moral philosophy is concerned with the values the individual ought to adopt and the rules the individual ought to observe. And, political philosophy is interested with the nature of the good, just, or true society and the problems regarding how public life ought to be organized. Furthermore, I was told that the chronology of ideas concerning philosophy could be seen as alternative possessions of truth or reality. While I have long abandoned the artificial dichotomy of moral and political philosophy because all ethical theories have practical imperatives that can be related to politics, I question whether the history of political theory is really about competing possessions of truth or reality.

Instead, I wonder if the history of political theory is not just about alternative perceptions of reality. One cannot, for example, reasonably claim that American politics can be explained and informed only by Burke's conservatism or Dewey's liberalism; rather, both can only offer discursive realities, because neither one possesses any universal truth. Similarly, John Locke's *Second Treatise of Government* or Karl Marx's *Communist Manifesto* are not political truths as much as they are reasonable versions of political realities in that one proclaims the truth of limited government and the other the validity of boundless socialism. As such, in this book I maintain the serviceability of political theory within the examination of environmental public policy because each philosophical version of risk analysis implies a view of the human condition that can help American citizens, bureaucrats, and academics understand the social life of risk as it is defined by the current modern liberalism. That being said, however, I contemplate whether my obsession for a well-organized anthology that represents a large scope of political realities with no absolute truth has biased my study.

As a consequence of reading risk analysis across the aforementioned canonical lenses of political liberalism rather than through a single comprehensive theorist, I speculate if I am giving an imprecise or sketchy view of risk analysis. Rather than arguing that, say, Jean Jacques Rousseau is the underpinning philosopher behind environmental risk policy in America, I have argued that Bentham, Rawls, and Habermas guide, but do not fully inform, risk assessment, management, and communication, respectively. Furthermore, I have painfully stated that other political thinkers should, therefore, augment each primary philosopher in order to repair the present impoverished form of modern liberalism used in contemporary risk analysis. According to some observers, this theoretical enterprise seems to get needlessly awkward because I suggest, for instance, that risk assessments are based on rational choice methodologies, which are a distorted form of preference seeking utilitarianism. I then argue at length that Bentham's project should be transformed with both John Stuart Mill's thought and Sandra Harding's criticism. Indeed, I am reminded that I was once told at a conference that while my conclusions about risk philosophy were interesting, they originated from a network of grand theorizing that would have no real applicability to actual risk policy.

The strongest criticism of this book is that because its research is not fluid, its analysis is not comprehensive and its findings are not relevant. A critical summary of my work might be as follows: Chapter Two discusses post-positive policy designs, Chapter Three argues for feminist epistemology, Chapter Four critiques social contract democracy, Chapter Five calls for a pluralistic understanding of environmental radicals, and Chapter Six relates discourse ethics with environmental communication. As referenced in the introduction, I would respond that my projected ethical matrix for risk analysis and my concluding template of the values found in the process of risk analysis is deliberately broken up and my analysis is both comprehensive and relevant because I examine liberalism as such. To the criticism that my work is not fluid, I would simply assert that my liberal analysis of risk is choppy because the rational practice of risk is divided as such.

In fact, the possible but not actual import of this book is that risk analysis can be more integrated if scientific experts of risk and social opponents of rationality both admit they have the same liberal ideas about political risk. Because of the multiple perceptions and judgments involved in risk analysis, it is sketchy by nature and attempts to examine it with a single lens will create a problem as legitimate as the one I started to address. Furthermore, this research offers a comprehensive descriptive commentary on risk analysis—and maybe a few prescriptive lessons—because, even though I am sampling three very different kinds of liberalism, they are not optimally chosen ideologies from the cannon of political philosophy, but the three theories that are currently informing risk practice.

Instead of providing a dynamic philosophy of post-modern risk analysis, I have created a reflexive modern liberalism that can comprehensively confront current risk debates because it is spawned from such language. In the essay "The Civilization of the Gene: Biotechnological Risk Framed in the Responsibility Discourse," it is argued that although debates over the contested use of biotechnology appear to be very exclusive, it is the most inclusive, broad, and comprehensive political topic to be debated. As explained in this book, topics about biotechnology reach every facet of modern civilization—social movements, religion, science, government, and industry. As with any other collective macro-political problem, any discussion about the exchanges, conflicts, and negations within that all encompassing institutional infrastructure will appear to be incomplete because agreement has not yet been reached on how to characterize the "risk society," "neo-corporatism," or "post-industrialism" (Strydom, 1999). The rebuttal to the accusation that my analysis of rBST and modern liberalism does not resolve the complex problems associated with biotechnology is that it did not purport to do so; rather it is a representative sample of the permanent discourse needed for Habermas's "complicated social processes of cooperation, conflict resolution, collective opinion and will formation"(Strydom, 1999, 28).

Finally, because the current risk practices and, by association, the political theories that guide them, are perceived as choppy and incomplete, some may question the relevance of this study. That is, even if one was to venture across the conventional ideas this book provides as it jumps from one topic of democracy to another, one must question: Why should anyone read an unsystematic theory of environmental liberalism that admits it cannot adequately explain all social phenomena surrounding risk analysis or resolve all the policy problems of food biotechnology? My argument has been that because of the new developments in biotechnology, a new version of modern liberalism needs to be advanced; and it should be one that both involves the rerationalization of probabilistic science and suggests reasonableness to the risk perceptions of society. "In this process, science's monopoly over the definition of what should count as scientific rationality is broken and many voices emerge around the theme of social responsibility. Accommodating these differences would appear to require a more inclusive and participatory democracy," both among policy experts and concerned citizens because it would be a modern liberalism that is concerned with a politics of respect instead of a policy of separation (Delanty, 1999, 37).

What I have proposed, therefore, is not a new environmental ethics or biopolitics distinct from all other varieties of practical knowledge, but an inquiry that reshapes, reforms, and re-evaluates decisionmaking processes in American environmental politics. As such, while ambitious, my proposed paradigm of modern liberalism with ecological sensibilities is relevant. Such liberalism not only entails rethinking our relationship to the environment, but also includes reconsidering the values and facts, the needs and attachments, and the actions and inactions of all political relationships. "There can be no universal or complete justification of any particular framework because it is not possible for us to anticipate all of the questions we will face nor the ways in which people will come to define themselves and their attachments and so come to discover norms..."(Moon, 1991, 224). Individuals should, therefore, come to risk political discourses with an open mind for if science is uncertain, then decisionmakers must be cautious and ought not rule on the political acceptability of a contested technology until all issues are addressed (Cooper, 1998). The current status of risk analysis as it applies to biotechnology suggests that such an open-minded notion of political liberalism and environmental ethics is not proceeding.

Almost all of the democratic theorists I endorse in this book advocate a strategy of progressive inclusion of as many new and different individuals and groups as possible. That is to say, some of theorists that I use—like Hobbes, Marcuse, or Young—do have aspects of their politics that are incompatible with liberalism, however, they all have a fundamental view about "what counts when we evaluate alternative political and social institutions and practices" (Schaeffer, 2001, 706). For example, when my book takes ideas from Robert Dahl, Carole Pateman, and Ulrich Beck, it might appear that I am constructing an irrelevant liberalism among radically dissimilar authors when, in fact, I am not. Dahl, a mid-century pluralist with corporatist tendencies, emphasized that a democracy is best served "by different experiences producing different interests, which should then be pursued by interest organization through groups"(Dryzek, 1996b, 477). Similarly, although she negates contractual liberalism, Pateman, as a feminist, is attuned to the need for actively inclusive representation of disadvantaged groups in order to change "universal rights and principles that seem to be tailored to the interests of upper-class white males" (Dryzek, 1996b, 476). Finally, "in the face of widespread public mistrust of arrogant scientists and technologists and of their corporate and governmental employers, Beck believes that legitimation in risk society can only be achieved by public participation in risk selection, allocation and amelioration"(Dryzek, 1996b, 480). Therefore, all of these thinkers would accept my idea of a modern liberalism that wishes to have dominant groups in public policymaking realize that their democracy has significant obstacles, limitations, and deficiencies for the new challenges offered by risk analysis and food biotechnology.

Finally, it should be noted that the attempt to truly know both the nature of political things and the right, good, or just political order is achieved not by interpreting a single political theory, but only by critically examining the text and the preferences of multiple political authors. Furthermore, my focus on the development of a discipline that desires order while distrusting all of the contributors of that order is not completely discredited because, as argued in this book, thought does not need to be completely rational in order to be reasonably considered. This is important because rather than possessing the truth, political philosophy is a process that serves to modify, control, and influence our perceptions of the truth in political life. For it has been said that science determines risk and population perceives risk, "both views are 'true,' but their truth consists in how risk beliefs support or challenge a cosmological framework of meaning and an established social order" (P. Thompson, 1999,7). With that perspective in mind, I think the idea of a theoretical case for reasonable environmental liberalism is relevant to scholars, citizens, and experts who are concerned about risk.

REFERENCES

POLITICAL THEORIES OF RISK ANALYSIS

Abbey, E. (1990). *The Monkey Wrench Gang*. Utah: Dream Garden Press.

Abbott, P. (1985). Henry David Thoreau, The State of Nature, and the Redemption of Liberalism. *The Journal of Politics, 47*(1), 182-208.

Advantage. (2001, July 11). A World of Biotechnology Benefits. *Monsanto Listserve*, from advantage@monsanto.com.

Alejandro, R. (1996). What is Political about Rawls's Political Liberalism? *The Journal of Politics, 58*(1), 1-24.

Arendt, H. (1958). *The Human Condition*. Chicago: Chicago University Press.

Arendt, H. (1972). *Crisis of the Republic*. New York: Harcourt Brace Jovanovich.

Armour, A. (1997). Rethinking the Role of Risk Assessment in Environmental Policymaking. In R. Bartlett & L. Caldwell (Eds.), *Environmental Policy: Transnational Issues and National Trends*. Westport: Quorum Books.

ARS-USDA. (2002, July 14). *Research Q&A: Bt Corn and Monarch Butterflies*, 2003, from www.ars.usda.gov/is/br/btcorn

Ball, T. (1973). Civil Disobedience and Civil Defiance. In R. B. Ripley (Ed.), *American Politics Series*. Beverly Hills: Sage.

Barber, B. (1984). *Strong Democracy: Participatory Politics for a New Age*. Berkeley: University of California Press.

Barber, B. (1994). Theory and Practice: Democracy and the Philosophers (2,500 Years of Democracy). *History Today, 44*(8), 44-50.

Barboza, D. (1999a, November 19, 1999). 2 Sides Square Off on Genetically Altered Food. *New York Times*.

Barboza, D. (1999b, November 12, 1999). Biotech Companies Take On Critics of Gene-Altered Food. *New York Times*.

Barboza, D. (2001, June 10, 2001). As Biotech Crops Multiply, Consumers Get Little Choice. *New York Times*.

Barnes, R. C., & Nowak, P. J. (1990). Bovine Somatotropin's Scale Neutrality and Constraints to Adoption. In S. M. Gendel, A. D. Kline, D. M. Warren & F. Yates (Eds.), *Agricultural Bioethics: Implications of Agricultural Biotechnology*. Ames: Iowa State University Press.

Barns, I. (1996). Environment, Democracy and Community. In F. Mathews (Ed.), *Ecology and Democracy*. London: Frank Cass and Company.

Barry, B. (1989). *Theories of Justice*. Berkeley: University of California Press.

Bartlett, R. (1986). Ecological Rationality: Reason and Environmental Policy. *Environmental Ethics, 8*(fall), 221-223.

Bartlett, R. (1998). Rationality and the Logic of the National Policy Act. In J. Dryzek & J. Schlosberg (Eds.), *Debating the Earth: The Environmental Politics Reader*. New York: Oxford University Press.

Bartlett, R., & Baber, W. (1999). From Rationality to Reasonableness in Environmental Administration: Moving Beyond Proverbs. *Journal of Management History, 5*(2), 55-67.

Bartlett, R. V., & Baber, W. F. (2001). Toward Environmental Democracy. *The Kansas Journal of Law and Public Policy, 11*(1), 35-64.

Baxter, T. (1997a). Environmental Justice for All. *Sierra, 82*(2), 101.

Baxter, T. (1997b). Making the Ties that Win. *Sierra, 82*(1), 54.

Beaudoin, K. (1999). Comment: On Tonight's Menu: Toasted Cornbread with Firefly Genes? Adapting Food Labeling Law to Consumer Protection Needs in the Biotech Century. *Marquette Law Review, 83*(Fall), 237-278.

Beck, U. (1992). *Risk Society: Towards a New Modernity*. London: Sage Publications.

Beck, U. (1994). Reinvention of Politics. In U. Beck, A. Giddens & S. Lash (Eds.), *Reflexive Modernity*. Stanford: Stanford University Press.

Beck, U. (1995). *Ecological Politics in an Age of Risk*. Cambridge: Polity Press.

Beck, U. (1997). The World as Laboratory. In *Twentieth Century Political Theory*. New York: Routledge.

Beckerman, W. (1999). Sustainable Development and Our Obligations to Future Generations. In A. Dobson (Ed.), *Fairness and Futurity*. Oxford: Oxford University Press.

Bedau, H. A. (1969). *Civil Disobedience: Theory and Practice*. New York: Macmillian Publishing Company.

Bentham, J. (1967). *Fragment on Government and An Introduction to the Principles of Morals and Legislation*. Oxford: Basil Blackwell.

Berman, L., & Murphy, B. A. (1999). *Approaching Democracy*. Upper Saddle River, NJ: Prentice Hall.

Bernstein, R. (1995). *The Restructuring of Social and Political Theory*. Philadelphia: University of Pennsylvania Press.

Bernstein, R. J. (Ed.). (1985). *Habermas and Modernity*. Cambridge: The MIT Press.

Berry, J. (1977). *Lobbying for the People*. Princeton: Princeton University Press.

Biro, A. (2000). *Re-Reading Marcuse in an Age of Technological Limits*. Paper presented at the Midwest Political Science Association Meeting, Chicago.

Bohman, J. (1996). *Public Deliberation: Pluralism, Complexity and Democracy*. Boston: MIT Press.

Bookchin, M., & Foreman, D. (1991). *Defending the Earth*. Boston: South End Press.

Boroush, M. (1998). *Understanding Risk Analysis: Internet Edition*. Washington, D.C.: American Chemical Society and Resources for the Future.

Brody, J. (2000, December 5, 2000). Gene Altered Foods: A Case Against Panic. *New York Times*.

Browne, W. P., & Hamm, L. G. (1990). Political Choices, Social Values, and the Economics of Biotechnology: A Lesson from the Dairy Industry. In D. J. Webber (Ed.), *Biotechnology: Assessing Social Impacts and Policy Implications*. New York: Greenwood Press.

Browne, W. P., Skees, J. R., Louis, S. E., Thompson, P. B., & Unnevehr, L. J. (1992). *Sacred Cows and Hot Potatoes*. Boulder: Westview Press.

Brulle, R. (2000). *Agency, Democracy and Nature: The U.S. Environmental Movement from a Critical Theory Perspective*. Cambridge, MA: The MIT Press.

Bullard, R. (1993). *Confronting Environmental Racism: Voices from the Grassroots*. Boston: South End Press.

Bultena, G., & Lasley, P. (1990). The Dark Side of Agricultural Biotechnology: Farmers' Appraisals of the Benefits and Costs of Technological Innovation. In S. Gendel, D. A. Kline, M. D. Warren & F. Yates (Eds.), *Agricultural Bioethics*. Ames: Iowa State University Press.

Burros, M. (1999a, December 1, 1999). Documents Show Officials Disagreed On Altered Food. *New York Times*.

Burros, M. (1999b, September 8, 1999). Eating Well; Different Genes, Same Old Label. *New York Times*.

Burton, J., & McBride, B. (1989a). Recombinant Bovine Somatotropin (rbST): Is There a Limit for Biotechnology in Applied Animal Agriculture. *Journal of Agricultural Ethics, 2*, 129-159.

Burton, J., & McBride, B. (1989b). Recombinant Bovine Somatotropin (rBST): Is There a Limit for Biotechnology in Applied Animal Agriculture? *Journal of Agricultural and Environmental Ethics, 2*(2), 129, 130.

Butler, L. J., & Cohn, G. (1993). The Economics of New Technologies in Dairying: BGH vs. Rotational Grazing. In W. C. Liebhardt (Ed.), *The Dairy Debate: Consequences of Bovine Growth Hormone and Rotational Grazing Technologies*. Davis: University of California Printing Services.

Bykerk, L., & Maney, A. (1991). Where Have All the Consumers Gone? *Political Science Quarterly, 8*(4), 677.

Byrd, D. M., & Cothern, R. (2000). *Risk Analysis: A Systematic Approach to Science-Based Decision Making*. Rockville, Maryland: Governmental Institutes.

Byrnes, K. (1993). Synthesis: Choices/Consequences. In W. Liebhardt (Ed.), *The Dairy Debate: Consequences of Bovine Growth Hormone and Rotational Grazing Techniques*. Davis: University of California Davis Press.

Carter, J. (1998, August 26). *Who's Afraid of Genetic Engineering?*, 2000, from www.biotechknowledge.com

Ceaser, J. W. (1990). *Liberal Democracy and Political Science*. Baltimore: John Hopkins University Press.
CEC. (2000). *White Paper on Food Safety*. Brussels: Commission of the European Communities.
Chalk, R. (1987). Looking for the Common Good. *Science, Technology and Human Values, 12*(3 and 4), 11-12.
Charles, D. (2001). *Lords of the Harvest: Biotech, Big Money and the Future of Food*. Cambridge: Perseus Publishing.
Charney, E. (1995). Political Liberalism, Deliberative Democracy, and the Public Sphere. *American Political Science Review, 92*(1), 97-101.
Cockburn, A. (1996, May 6). EARTH FIRST!, the Press and the Unabomber. *The Nation, 262*, 6-7.
Code, L. (1993). Taking Subjectivity into Account. In *Feminist Epistemologies*. New York: Routledge.
Cohen, J. L., & Arato, A. (1993). *Civil Society and Political Theory*. Cambridge: MIT Press.
Comstock, G. (1990). The Case Against bGH. In S. M. Glendel, D. A. Kline, M. D. Warren & F. Yates (Eds.), *Agricultural Bioethics: Implications of Agricultural Biotechnology*. Ames: Iowa University Press.
Cooper, G. (1998). Teleology and Environmental Ethics. *American Philosophical Quarterly, 35*(2), 195-208.
Council, N. R. (1983). *Risk Assessment in the Federal Government: Managing the Process*. Washington, D.C.: National Academy Press.
Cox, R. J. (1996). Making Democracy Work. *Sierra, 81*, 51.
Cranor, C. F. (1990). Some Moral Issues in Risk Assessment. *Ethics, 101*(1), 123-143.
Crosby, A. W. (1995). The Past and Present of Environmental History. *The American Historical Review, 100*(4), 1177-1189.
CRS. (1999). *Food Biotechnology in the United States: Science, Regulation, and Issues*. Congressional Research Service: June 02.
Curtis, M. (1961a). *The Great Political Theories: Volume One*. New York: Avon Books.
Curtis, M. (1961b). *The Great Political Theories: Volume Two*. New York: Avon.
Cutter, S. (1993). *Living With Risk*. New York: Routledge.
Dahl, R. (1956). *A Preface to Democratic Theory*. Chicago: Chicago University Press.
Dahl, R. (1983). *Dilemmas of Pluralist Democracy*. New Haven: Yale University Press.
Dahl, R. (1989). *Democracy and Its Critics*. New Haven: Yale University Press.
deGrazia, V. (1996). *The Sex of Things: Gender and Consumption in Historical Practice*. Berkeley: University of California Press.
Deibert, A. C. (1990). Agricultural Biotechnology: Public Perception of Risk. In S. M. Gendel, A. D. Kline, D. M. Warren & F. Yates (Eds.), *Agricultural Bioethics: Implications of Agricultural Biotechnology*. Ames: Iowa State University Press.
Deitz, M. (1998). Context Is All: Feminism and Theories of Citizenship. In A. Phillips (Ed.), *Feminism and Politics*. Oxford: Oxford University Press.
Delanty, G. (1999). Biopolitics in the Risk Society: The Possibility of a Global Ethic of Social Responsibility. In P. O'Mahony (Ed.), *Nature, Risk and Responsibility*. New York: Routledge.
deLeon, P. (1992). The Democratization of the Policy Sciences. *Public Administration Review, 52*(1), 125-129.
deLeon, P. (1994). Reinventing the Policy Sciences: Three Steps Back into the Future. *Policy Sciences, 52*(1), 77-95.
deLeon, P. (1995). Democratic Values and the Policy Sciences. *American Journal of Political Science, 39*(4), 886-905.
de-Shalt, A. (2000). *The Environment: Between Theory and Practice*. Oxford: Oxford University Press.
Dobson, A. (1998). *Justice and the Environment*. Oxford: Oxford University Press.
Dobson, A. (1999a). *Fairness and Futurity*. Oxford: Oxford University Press.
Dobson, A. (1999b). *Justice and the Environment*. Oxford: Oxford University Press.
Donelan, J. E. (1995). *Civil Disobedience, Civil Society, and the State*. Unpublished Ph.D., State University of New York at Stony Brook, Stony Brook.
Douglas, M., & Wildavsky, A. (1982). *Risk and Culture*. Berkeley: University of California Press.
Douthitt, R., & Mayer, R. (1995). Consumer Risk Reception and Recombinant Bovine Growth Hormone: The Case for Labeling Dairy Products Made From Untreated Herd Milk. *Journal of Public Policy and Marketing, 14*(2), 328-332.
Dowie, M. (1995). *Losing Ground*. Cambridge: The MIT Press.

Doyle, J. (1985). *Altered Harvest: Agriculture, Genetics, and the Fate of the World's Food Supply*. New York: Viking.
Doyle, J. (1990). Who Will Gain from Biotechnology. In S. M. Gendel, A. D. Kline, D. M. Warren & F. Yates (Eds.), *Agricultural Bioethics: Implications of Agricultural Biotechnology*. Ames: Iowa State University.
Drury, S. (1988). *The Political Ideas of Leo Strauss*. New York: St. Martin's Press.
Dryzek, J. (1990). *Discursive Democracy*. Cambridge: Cambridge University Press.
Dryzek, J. (1996a). Political and Ecological Communication. In F. Matthews (Ed.), *Ecology and Democracy*. London: Frank Cass.
Dryzek, J. (1996b). Political Inclusion and the Dynamics of Democratization. *The American Political Science Review, 90*(3), 475-487.
Dryzek, J. (1997). *The Politics of the Earth: Environmental Discourses*. New York: Oxford University Press.
Dryzek, J. (2000). Justice and the Environment: Concepts of Environmental Sustainability and Dimensions of Social Justice and Capitalism, Democracy, and Ecology: Departing from Marx and Environmentalism for a New Millennium: The Challenge of Coevolution, book review. *American Political Science Association, 94*(1), 169-173.
Dunn, J. (1985). Trust in the Politics of John Locke. In *Rethinking Modern Political Theory*. Cambridge: Cambridge University Press.
Dunn, W. (1983). *Values, Ethics and the Practice of Policy Analysis*. Lexington, MA: Lexington Books.
Durant, R. (1995). The Democratic Deficit in America. *Political Science Quarterly, 110*(1), 25-47.
Dye, T. R. (1992). *Understanding Public Policy* (7th ed.). Englewood Cliffs, NJ: Prentice Hall.
Eagan, S. (1996). From Spikes to Bombs: The Rise of Eco-Terrorism. *Studies in Conflict and Terrorism, 19*(1), 1-18.
Easterbrook, G. (1995). *A Moment on Earth: The Coming Age of Environmental Realism*. New York: Viking Press.
Echols, M. (1998). Food Safety Regulation in the European Union and the United States: Different Cultures, Different Laws. *Columbia Journal of European Law, 4*(Summer), 525-543.
Eckersley, R. (1992). *Environmentalism and Political Theory*. Albany: State University of New York Press.
Eichenwald, K., Kolata, G., & Petersen, M. (2001, January 26, 2001). Biotechnology Food: From the Lab to a Debacle. *New York Times*, p. C1;C6.
Enteman, W. (1993). *Managerialism: The Emergence of a New Ideology*. Madison: The University of Wisconsin Press.
ERS. (1990). *Biotechnology and Agriculture: Emergence of Bovine Somatotropin*. Economic Research Service: Commodity Economics Division.
Ewens, L. E. (2000). Seed Wars: Biotechnology, Intellectual Property, and the Quest for High Yield Seeds. *Boston College International and Comparative Law Review, 23*(Spring), 285-310.
Falcon, W. P. (2001). Sustainable Science for a Sustainable Environment: Comments on Searching for Sustainability. *Ecology Law Quarterly, 27*(1), 1163-1174.
FDA. (1998, August 12). *Policy for Foods Developed by Biotechnology: Education/Compliance Program 1995*, 2000, from http://vm.cfsan.fda.gov
FDA. (1999a, January 21). *Review of the Safety of Recombinant Bovine Somatotropin*, 2000, from http://www.fda.gov/cvm/fd/infores/other/RBRTFNL
FDA. (1999b, Febuary). *Update on Human Food Safety of BST*, 2001, from http://www.fda.gov/cvm/infores/updates/BSTSAFUP.html
Feder, B. (2000, January 1, 2000). Visions: Biology: Science Invades the Pantry. *New York Times*.
FederalRegister. (1992, 29 May). Statement of Policy: Foods Derived From New Plant Varieties, *57*, 22984-23005.
FederalRegister. (1993, 28 April). Food Labeling: Foods Derived from New Varieties, *58*, 25837-25841.
Feenberg, A. (1999). *Questioning Technology*. New York: Routledge.
Ferre, F., & Hartel, P. (1994). *Ethics and Environmental Policy: Theory Meets Practice*. Athens, GA: University of Georgia Press.
Fisch, M. (1951). *Classic American Philosophers: Peirce, James, Royce, Santayana, Dewey, Whitehead*. Englewood Cliffs: Prentice-Hall.
Fisher, F. (1990). *Technocracy and the Politics of Expertise*. London: Sage Publications.

Fisher, F., & Forester, J. (Eds.). (1993). *The Argumentative Turn in Policy Analysis and Planning*. Durham: Duke University Press.

FMI. (2001, December 1). *Genetically Modified Food and Biotechnology: A Public Policy Position*, 2001, from www.fmi.org

Foreman, D., & Haywook, B. (1989). *Ecodefense: A Field Guide to Monkeywrenching*. Tucson: Ned Ludd Books.

Fortas, A. (1972). Concerning Dissent and Civil Disobedience. In H. A. Bosmajian (Ed.), *Dissent: Symbolic Behavior and Rhetorical Strategies*. Boston: Allyan and Bacon.

Fox, M. D. (1997). *Eating with Conscience: The Bioethics of Food*. Troutdale, OR: New Sage Press.

Frazier, O. B. N. (2001, August 08, 2001). Is it Time to Slow Down and Smell the Genetically Altered Roses? *Catholic News Service*, pp. 1-2.

Friedman, J. (Ed.). (1996). *The Rational Choice Controversy: Economic Models of Politics Reconsidered*. New Haven: Yale University Press.

Galston, W. (1980). *Justice and the Human Good*. Chicago: University of Chicago Press.

GAO. (1992). *Recombinant Bovine Growth Hormone: FDA Approval Should be Withheld Until the Mastitis Issue is Resolved*. General Accounting Office: Summary Report to Congressional Requesters.

Gaskell, G., & Bauer, M. (Eds.). (2001). *Biotechnology 1996-2000*. London: Science Museum.

Gendel, S., Kline, A. D., Warren, D. M., & Yates, F. (Eds.). (1990). *Agricultural Bioethics: Implications of Agricultural Biotechnology*. Ames: Iowa State University Press.

Gendin, S. (1972). Government Toleration of Civil Disobedience. In V. Held, K. Nielsen & C. Parsons (Eds.), *Philosophy and Political Action*. London: Oxford University Press.

Gibbs, L. (1983). *Love Canal*. Albany: State University of New York Press.

Gillroy, J. M. (1993). Integrity, Value and Analysis of Risk. In J. M. Gilroy (Ed.), *Environmental Risk, Environmental Values and Political Choices*. Boulder: Westview Press, Inc.

Goldwin, R. (1969). *On Civil Disobedience: American Essays, Old and New*. Chicago: Rand McNally.

Goodin, R. (1995). *Utilitarianism as a Public Philosophy*. Cambridge: Cambridge University Press.

Goodin, R. E. (1982). *Political Theory and Public Policy*. Chicago: The University of Chicago Press.

Goss, P. J. (1996). Guiding the Hand that Feeds: Toward Socially Optimal Appropriability in Agricultural Biotechnology Innovation. *California Law Review, 84*(October), 1395-1436.

Gottlieb, R. (1993). *Forcing the Spring: The Transformation of the American Environmental Movement*. Washington, D.C.: Island Press.

Graham, J. D. (Ed.). (1995). *Risk vs. Risk: Tradeoffs in Protecting Health and the Environment*. Cambridge: Harvard University Press.

Grobe, D., & Douthitt, R. (1996). A Model of Consumers' Risk Perceptions Toward Recombinant Bovine Growth Hormone (RBGH): The Impact of Risk Characteristics. *Consumer Interests Annual*(42), 395-401.

Grobe, D., & Douthitt, R. (1999). Consumer Risk Perception Profiles Regarding Recombinant Bovine Growth Hormone (RBGH). *Journal of Consumer Affairs, 33*(2), 254-276.

Gunnell, J. (1993). American Political Science, Liberalism, and the Invention of Political Theory. In J. Farr & R. Seidelman (Eds.), *Discipline and History*. Ann Arbor: University of Michigan Press.

Habermas, J. (1968). *Toward a Rational Society: Student Protest, Science and Politics* (J. Shapiro, Trans.). Boston: Beacon Press.

Habermas, J. (1979). *Communication and the Evolution of Society* (T. McCarthy, Trans.). Boston: Beacon Press.

Habermas, J. (1987). *The Theory of Communicative Action: Volume Two, Lifeworld and System* (T. McCarthy, Trans.). Boston: Beacon Press.

Habermas, J. (1994). *The Past as Future* (M. Pensky, Trans.). Lincoln: The University of Nebraska Press.

Habermas, J. (1996). *Between Facts and Norms: Contributions to a Discourse Theory of Law and Democracy* (W. Rehg, Trans.). Cambridge: MIT Press.

Hajer, M. (1995). *The Politics of Environmental Discourse: Ecological Modernization and the Policy Process*. Oxford: Oxford University Press.

Hall, R. T. (1971). *The Morality of Civil Disobedience*. New York: Harper and Row.

Handmer, J. W., Norton, T. W., & Dovers, S. R. (2001). *Ecology, Uncertainty, and Policy*. Essex: Pearson Education Limited, Prentice Hall.

Hardin, R. (1988). *Morality within the Limits of Reason*. Chicago: Chicago University Press.

Hardin, R. (1991). Rereading Hobbes in other Words: Contractarian, Utilitarian, Game Theorist. *Political Theory, 19*(2), 156-180.
Harding, S. (1983). Why Has the Sex/Gender System Become Visible Only Now? In S. Harding & M. B. Hintikka (Eds.), *Discovering Reality: Feminist Perspectives on Epistemology, Metaphysics, Methodology, and Philosophy of Science*. London: D. Reidel Publishing Company.
Harding, S. (1986). *The Science Question in Feminism*. Ithaca, NY: Cornell University Press.
Harding, S. (1993). Rethinking Standpoint Epistemology: "What is Strong Objectivity?" In L. Alcoff & E. Potter (Eds.), *Feminist Epistemologies*. New York: Routledge.
Harsanyi, J. C. (1975). Can the Maximin Principle Serve as a Basis for Morality? A Critique of John Rawls' Theory. *American Political Science Association, 59*(2), 594-605.
Hartsock, N. C. M. (1998). *The Feminist Standpoint Revisited and Other Essays*. Boulder: Westview Publishing.
Hartz, L. (1955). *The Liberal Tradition in America*. New York: Harcourt Brace and Company.
Hayry, M. (2000). How to Apply Ethical Principles to the Biotechnological Production of Food: The Case of Bovine Growth Hormone. *Journal of Agricultural and Environmental Ethics, 12*, 177-184.
HCPD. (1998, 10 October). *Overview of Review of Recombinant Bovine Somatotropin (rBST) by Health Canada*, 2000, from http://www.hc-sc.gc.ca/english/archivesrelease
Held, V. (1971). Civil Disobedience and Public Policy. In E. Kent (Ed.), *Revolution and the Rule of Law*. Englewood Cliffs, NJ: Prentice.
Held, V. (1972). On Understanding Political Strikes. In V. Held, K. Nielsen & C. Parsons (Eds.), *Philosophy and Political Action*. New York: Oxford University Press.
Held, V. (1985). Feminism and Epistemology: Recent Work on the Connection between Gender and Knowledge. *Philosophy and Public Affairs, 14*(3), 296-307.
Held, V. (1993). *Feminist Morality: Transforming Culture, Society, and Politics*. Chicago: University of Chicago.
Henderson, D. (1993). Science, Environment and Policy Prescriptions. In J. M. Gillroy (Ed.), *Environmental Risk, Environmental Values, and Political Choices: Beyond Efficiency Trade-offs in Public Policy Analysis*. Boulder: West View.
Hettinger, N. (1995). Patenting Life: Biotechnology, Intellectual Property, and Environmental Ethics. *Boston College Environmental Affairs Law Review, 22*(Winter), 267-305.
Hirschmann, N. J. (1989). Freedom, Recognition, and Obligation: A Feminist Approach to Political Theory. *American Political Science Review, 83*(4), 1227-1244.
Hobbes, T. (1994). *Leviathan, 1651* (C. B. Machpherson, Trans.). Indianapolis: Hackett Publishing.
Hood, C., Rothstein, H., & Baldwin, R. (2001). *The Government of Risk: Understanding Risk Regulation Regimes*. Oxford: Oxford University Press.
Horning, S. (1993). Reading Risk: Public Response to Print Media Accounts of Technological Risk. *Public Understanding of Science, 2*(1), 95-109.
Hornstein, D. T. (1995). Paradigms, Process, and Politics: Risk and Regulatory Design. In A. Finkel & D. Golding (Eds.). Washington, D.C.: Resources for the Future.
IFST. (1999, September 11). *Position Statement, IFST Current Hot Topics: Bovine Somatotropin*, 2001, from http://www.ifst.org/hottop8a.htm.
Ingalsbee, T. (1996). Earth First! Activism: Ecological Postmodern Praxis in Radical Environmental Identities. *Sociological Perspectives, 39*(2), 241-263.
Ingram, H., & Rathgeb, S. (Eds.). (1995). *Public Policy for Democracy*. Washington, D.C.: The Brookings Institution.
InternalReview. (1998). *rBST Gaps Analysis*. Ottawa: Health Protection Branch, Health Canada.
Jackson, W., & Piper, J. (1989). The Necessary Marriage Between Ecology and Agriculture. *Ecology, 70*(6), 1591-1593.
Johnson, J. (1991). Habermas on Strategic and Communicative Action. *Political Theory, 19*(2), 181-201.
Juskevich, J. C., & Guyer, C. G. (1990). Bovine Somatotropin: Review of an Emerging Animal Technology. *Science, 249*(84), 875.
Kalter, R. (1985). The New Biotech Agriculture: Unforeseen Economic Consequences. *Issues in Science and Technology, 13*(1), 125-133.
Kasperson, R., Golding, D., & Tuler, S. (1992). Social Distrust as a Factor in Siting Hazardous Facilities and Communicating Risk. *Journal of Social Issues, 48*(4), 161-187.

Kaufman, M. (2001, March 26-April 1, 2001). Stalking Corn That Can Make You Sick. *Washington Post National Weekly Review*, p. 29.
Kellner, D. (1991). Introduction. In *One-Dimensional Man*. Boston: Beacon Press.
Keulartz, J., Korthals, M., Schrmer, M., & Swierstra, T. (2002). *Pragmatist Ethics for a Technological Culture*. Dordrecht, The Netherlands: Kluwer Academic Publishers.
Kneen, B. (1999). *Farmageddon: Food and the Culture of Biotechnology*. Gabriola, British Columbia, Canada: New Society Publishers.
Kondro, W. (2000). Canadian Scientists Vindicated Over Whistleblowing. *The Lancet, 356*(9235), 1091.
Kraft, M. (1996). *Environmental Politics and Policy: Toward the Twenty-First Century*. New York: HarperCollins.
Kraft, M., & Vig, N. (Eds.). (1994). *Environmental Policy in the 1990's* (2nd ed.). Washington, D.C.: Congressional Quarterly Press.
Krimsky, S. (1991). *Biotechnics and Society: The Rise of Industrial Genetics*. New York: Praeger.
Krimsky, S. (1999). Symposium: The Profit of Scientific Discovery and Its Normative Implications. *Chicago-Kent College of Law Review, 75*(15), 1-20.
Krimsky, S., & Plough, A. (1988). *Environmental Hazards: Communicating Risks as a Social Process*. Dover, MA: Auburn House Publishing Company.
Krutch, J. W. (1962). *Walden and Other Writings of Henry David Thoreau*. New York: Bantam Books.
Kuchler, F., McClelland, J., & Offutt, S. E. (1990). Regulatory Experience with Food Safety: Social Choice Implications for Recombinant DNA-Derived Animal Growth Hormones. In D. Webber (Ed.), *Biotechnology: Assessing Social Impacts and Policy Implications*. New York: Greenwood Press.
Kukathas, C., & Pettit, P. (1990). *Rawls: A Theory of Justice and its Critics*. Stanford: Stanford University Press.
Kymlicka, W. (1989). Liberal Individualism and Liberal Neutrality. *Ethics, 99*(4), 883-905.
Laird, F. (1993). Participatory Analysis, Democracy, and Technological Decision Making. *Science, Technology and Human Values, 18*(3), 341-361.
LawReporter. (1996). Corporate Rights, and Wrongs. *Multinational Monitor, 17*, 5-6.
Lebacqz, K. (1986). *Perspectives from Philosophical and Theological Ethics: Six Theories of Justice*. Minneapolis: Augsburg Publishing House.
Lee, M. F. (1995). *Earth First!* Syracuse: Syracuse University Press.
Lehman, H. (1995). *Rationality and Ethics in Agriculture*. Moscow, ID: University of Idaho Press.
Lester, J. P. (Ed.). (1995). *Environmental Politics and Policy*. Durham: Duke University Press.
Liebhardt, W. (1993). *The Dairy Debate: Consequences of Bovine Growth Hormone and Rotational Grazing Technologies*. Davis: University of California Printing Services.
Light, A., & Katz, E. (Eds.). (1996). *Environmental Pragmatism*. New York: Routledge.
Lineberry, R. L., & Edwards, G. C. I. (1989). *Government in Action* (Fourth ed.). Boston: Scott, Foresman and Company.
List, P. (Ed.). (1993). *Radical Environmentalism*. Belmont: Wadsworth Publishers.
Locke, J. (1993). The Second Treatise of Government, 1681. In D. Wooton (Ed.), *Political Writings of John Locke*. New York City: Penguin Books.
Long, N. E. (1949). Power and Administration. *Public Administration Review, 9*, 257-264.
Lowi, T. J. (1979). *The End of Liberalism*. New York: W.W. Norton Company.
Luke, T. (1999). *Capitalism, Democracy, and Ecology*. Champaign: University of Illinois Press.
Lukes, T. (1994). Marcuse and Ecology. In J. Bokina & T. Lukes (Eds.), *Marcuse, From the New Left to the Next Left*. Lawrence: Kansas University Press.
MacKinnon, C. (1982). Feminism, Marxism, Method and State: An Agenda for Theory. *Signs, 7*(3), 1-24.
Maclean, A. (1993). *The Elimination of Morality: Reflections on Utilitarianism and Bioethics*. New York: Routledge.
MacLean, D. (1986). Philosophical Issues for Centralized Decisions. In D. MacLean (Ed.), *Values at Risk*. Totawa, NJ: Rowman & Allanheld.
Majone, G. (1989). *Evidence, Argument, and Persuasion in the Policy Process*. New Haven: Yale University Press.
Malkin, M. (2001, May 29, 2001). Violent Acts in the Name of the Environment. *The Indianapolis Star*, p. 8.

Manning, R. (1981). Environmental Ethics and Rawls' Theory of Justice. *Environmental Ethics, 3*, 155-165.
Mansbridge, J. (1980). *Beyond Adversary Democracy*. New York: Basic Books Inc.
Mansbridge, J. (1993). Feminism and Democratic Community. In J. Chapman & I. Shapiro (Eds.), *Nomos XXXV: Democratic Community*. New York: New York University Press.
Marcuse, H. (1964). *One-Dimensional Man*. Boston: Beacon Press.
Marcuse, H. (1966). *Eros and Civilization*. Boston: Beacon Press.
Marcuse, H. (1972). *Essay on Liberation*. London: Penguin Books.
Margolis, H. (1996). *Dealing with Risk: Why the Public and Experts Disagree on Environmental Issues*. Chicago: University of Chicago Press.
Marshall, C., & Rossman, G. (Eds.). (1989). *Designing Qualitative Research*. London: Sage Publications.
Martineau, B. (2001). Food Fight. *The Sciences, 41*(2), 24-36.
Matavalli, J., & Rembert, T. (1996). A Brave New World: Genetic Engineering, RBGH and Mad Cow Disease. *E Magazine: The Environmental Magazine, 7*(40), 2-4.
Mathews, F. (Ed.). (1996). *Ecology and Democracy*. London: Frank Cass and Company.
Mayo, D. (1997). Sociological Versus Metascientific Views of Technological Risk Assessment. In K. Shrader-Frechette & L. Westra (Eds.), *Technology and Values*. Lanham: Rowman and Littlefield Publishers.
McCarthy, T. (1978). *The Critical Social Theory of Jurgen Habermas*. Cambridge: The MIT Press.
McCool, J. (Ed.). (1995). *Public Policy Theories, Models and Concepts*. Englewood Cliffs: Prentice Hall.
McDonald, F. (1985). *Novos Ordo Seclorum: Intellectual Origins of the Constitution*. Lawrence: Kansas University Press.
McGlouglin, M. (1999, November 11). *We'll Starve Without Biotech: Genetic Engineering is Subject to More Safeguards than Many Unaltered Foods We Eat*, 2001, from www.biotechknowledge.com
Mepham, B. (2000). A Framework for the Ethical Analysis of Novel Foods: The Ethical Matrix. *Journal of Agricultural and Environmental Ethics, 12*, 165-176.
Mepham, B. (Ed.). (1996). *Food Ethics*. London: Routledge.
Mephan, T. B., Tucker, G. A., & Wiseman, J. (Eds.). (1995). *Issues in Agricultural Bioethics*. Nottingham: Nottingham University Press.
Meredith, H., & Robertson, L. (2000). Getting a Hearing. *American Journalism Review, 22*(7), 16.
Mill, J. S. (1950). *On Bentham and Coleridge*. New York: Harper and Row.
Mill, J. S. (1962). *John Stuart Mill: Utilitarianism On Liberty; essay on Bentham*. Cleveland: Meridian.
Mill, J. S. (1988). *Utilitarianism, On Liberty and Considerations on Representative Government*. London: J.M. Dent and Sons.
Mills, W. C. (1956). *The Power Elite*. Oxford: Oxford University Press.
Mollin, M. (2000). *Actions Louder than Words: Gender and Activism in the American Radical Pacifist Movement, 1942-1972*. University of Massachusetts, Amhert.
Monsanto. (1998, December 18). *Medical Background Information*, 2000, from www.monsanto.com/mediacenter/background/98dec18_ABC.html
Monsanto. (2000, January 3). *Status Update Regarding: Posilac Bovine Somatotropin*, 2001, from www.monsanto.com/monsanto/mediacenter/background/posilac.html
Monsanto. (2001a, November 7). *The History of BST and its Manufacturing and Administration*, 2001, from www.monsanto.com/dairy8_m&moh.html
Monsanto. (2001b, July 14). *Status Update Regarding: Posilac Bovine Somatotropin*, 2001, from www.monsanto.com/monsanto/mediacenter/background/jul14_posilac.html
Monsanto. (2002, November 13). *The New Monsanto Pledge*, 2002, from www.monsanto.com
Moon, D. (1991). Constrained Discourse and Public Life. *Political Theory, 19*(2), 202-229.
Morgan, R. (Ed.). (1996). *Sisterhood is Global*. New York: The Feminist Press at the City University of New York.
Morris, K. (2000). Bovine-somatotropin controversy heads into Europe. *The Lancet, 353*(9159), 1165.
Murphy, R. (1994). *Rationality and Nature: A Sociological Inquiry into a Changing Relationship*. Boulder: Westview Press.
Nagel, J. (1975). *The Descriptive Analysis of Power*. New Haven: Yale University Press.
Naik, P. S. (2000). Biotechnology Through the Eyes of an Opponent: The Resistance of Activist Jeremy Rifkin. *Virginia Journal of Law and Technology, 5*(Spring), n1-n138.

Nash, R. F. (1989). *The Rights of Nature: A History of Environmental Ethics*. Madison: The University of Wisconsin Press.
Norton, B. (1999). Ecology and Opportunity: Intergenerational Equity and Sustainable Options. In A. Dobson (Ed.), *Fairness and Futurity*. Oxford: Oxford University Press.
Oakeshott, M. (1962). *Rationalism in Politics and Other Essays*. Indianapolis: Liberty Press.
OCA. (2003, August 01). *recombinant Growth Hormone*, 2003, from www.organicconsumers.org/
Ogden, C. K. (1993). *Jeremy Bentham 1832-2032*. London: Thoemmes Press.
Ophuls, W., & Boyan, S. A. J. (1992). *Ecology and the Politics of Scarcity Revisited*. New York: W.H. Freeman and Company.
Osborne, R. (1992). *Philosophy for Beginners*. New York: Writers and Readers Publishing.
Paehlke, R. (1990). Democracy and Environmentalism: Opening the Door to the Administrative State. In R. Paehlke & D. Torgerson (Eds.), *Managing the Leviathan: Environmental Politics and the Administrative State*. New York: Broadview Press.
Paehlke, R. (1994). Environmental Values and Public Policy. In N. J. Vig & M. E. Kraft (Eds.), *Environmental Policy in the 1990's* (2nd ed.). Washington, D.C.: Congressional Quarterly Press.
Parenti, M. (1995). *Democracy for the Few*. New York: St. Martin's Press.
Pateman, C. (1970). *Participation and Democratic Theory*. Cambridge: Cambridge University Press.
Pateman, C. (1985). *The Problem of Political Obligation*. Berkeley: University of California Press.
Pateman, C. (1988). *The Sexual Contract*. Stanford: Stanford University Press.
PEW. (2003). *Factsheet: Genetically Modified Crops in the United States*. Retrieved June 29, 2003, from http://pewagbiotech.org/
Pinkerton, J. (1997). Enviromanticism: The Poetry of Nature as a Political Force. *Foreign Affairs, 76*(3), 2-.
Pistorius, R., & Wifk, J. v. (1999). *The Exploitation of Plant Genetic Information: Political Strategies in Crop Development*. Cambridge: CAB International.
Plato. (1974). *The Republic, 385 B.C.* (G. M. A. Grube, Trans.). Indianapolis: Hackett Publishing Company.
Plough, A., & Krimsky, S. (1987). The Emergence of Risk Communication Studies: Social and Political Context. *Science, Technology and Human Values, 12*(3 and 4), 4-10.
Plumwood, V. (1996). Has Democracy Failed Ecology? A Feminist Perspective. In F. Mathews (Ed.), *Ecology and Democracy*. London: Frank Cass and Company.
Plutzer, E., Maney, A., & O'Connor, R. E. (1998). Ideology and Elites' Perceptions of the Safety of New Technologies. *American Journal of Political Science, 42*(1), 190-209.
Pojman, L. (1997). What is Moral Philosophy. In K. Shrader-Frechette & L. Westra (Eds.), *Technology and Values*. Lanham: Rowman and Littlefield Publishers.
Powell, D. (1999, May). Ottawa's Errant Food Fight. *National Post*, C7.
Powell, D., & Leiss, W. (Eds.). (1997). *Mad Cows and Mother's Milk: The Perils of Poor Risk Communication*. Montreal: McGill-Queen's University Press.
Power, P. F. (1970). On Civil Disobedience in Recent American Democratic Thought. *American Political Science Review, 64*(1), 35-47.
Press, D. (1994). *Democratic Dilemmas in the Age of Ecology*. Durham: Duke University Press.
Pritchard, M., & Robinson, W. (1981). Justice and the Treatment of Animals: A Critique of Rawls. *Environmental Ethics, 3*(55-61).
Quayle, D. (1991). *Report on National Biotechnology Policy*. Washington, D.C.: The President's Council on Competitiveness.
Ratzan, S. (Ed.). (1998). *The Mad Cow Crisis Health and the Public Good*. New York: New York University Press.
Rawls, J. (1971a). The Justification of Civil Disobedience. In E. Kent (Ed.), *Revolution and the Law*.
Rawls, J. (1971b). *A Theory of Justice*. Cambridge: Harvard University Press.
Rawls, J. (1992). Justice as Rational Choice Behind a Veil of Ignorance. In J. B. Sterba (Ed.), *Justice: Alternative Perspectives*. Belmont: Wadsworth Publishing Company.
Rawls, J. (1993). *Political Liberalism*. New York: Columbia University Press.
Rawls, J. (1995). Political Liberalism: Reply to Habermas. *The Journal of Philosophy, 92*(3), 132-180.
Rehg, W. (1994). *Insight and Solidarity: The Discourse Ethics of Jurgen Habermas*. Los Angeles: University of California.

Reiss, M., & Straughan, R. (1996). *Improving Nature: The Science and Ethics of Genetic Engineering*. New York: Cambridge.

Renn, O., Burns, W., Kasperson, J., Kasperson, R., & Slovic, P. (1992). The Social Amplification of Risk: Theoretical Foundations and Empirical Applications. *Journal of Social Issues, 48*(4), 137-160.

Reporter, A. P. (1997, August 16, 1997). Ben & Jerry's Scoops Monsanto: Hormone-Free Labeling Allowed. *St. Louis Dispatch*, p. B 25.

Rifkin, J., & Rifkin, C. G. (1992). *Voting Green*. New York: Doubleday.

Ring, J. (1991). *Modern Political Theory and Contemporary Feminism*. New York: State University of New York.

Robbins, S. P. (1997). *Essentials of Organizational Behavior* (5th ed.). Upper Saddle River, NJ: Prentice Hall.

Roosevelt, M. (2000, July 31, 2000). Taking It to Main Street. *Time, 7*, 42-43.

Rosen, R. (1994). Who Gets Polluted? The Movement for Environmental Justice. *Dissent, 17*, 223-230.

Rucker, D. (1966). The Moral Grounds of Civil Disobedience. *Ethics, 76*(2), 142-145.

Sandel, M. (Ed.). (1984). *Liberalism and Its Critics*. New York: New York University Press.

Sanders, L. (1997). Against Deliberation. *Political Theory, 25*(3), 347-376.

Sandman, P., Hance, B. J., & Chess, C. (1990). *Industry Risk Communication Manual: Improving Dialogue with Communities*. Boca Raton: Lewis.

Sartori, G. (1987). *The Theory of Democracy Revisited*. Chantham, N.J.: Chantham House.

Schaeffer, D. (2001). Feminism and Liberalism Reconsidered: The Case of Catharine MacKinnon. *American Political Science Review, 95*(3), 699-708.

Scott, J. C. (1990). *Domination and the Arts of Resistance: Hidden Transcripts*. New Haven: Yale University Press.

Seager, J. (1993). *Earth Follies: Coming to Feminist Terms with the Global Environmental Crisis*. New York: Routledge.

Sergerson, K., & Bromley, D. (1985). *The Social Response to Environmental Risk: Policy Formulation in an Age of Uncertainty*. Boston: Kluwer Academic.

Shapiro, R. B. (2000). *The Welcome Tension of Technology: The Need for Dialogue about Agricultural Biotechnology*. St. Louis: Center for the Study of American Business, Washington University.

Sherlock, R., & Kawar, A. (1990). Regulating Genetically Engineered Organisms: The Case of the Dairy Industry. In D. Webber (Ed.), *Biotechnology: Assessing Social Impacts and Policy Implications*. New York: Greenwood Press.

Shiva, V. (1994). *Staying Alive*. London: Zed Books.

Shrader-Frechette, K. (1985). *Science Policy, Ethics, and Economic Methodology: Some Problems of Technology Assessment and Environmental Impact Analysis*. Boston: D. Reidel.

Shrader-Frechette, K. (1989). Policy through Assessment of Type II Statistical Risks. In R. V. Bartlett (Ed.), *Policy Through Impact Assessment: Institutionalized Analysis As A Policy Strategy*. New York: Greenwood Press.

Shrader-Frechette, K. S. (1991). *Risk and Rationality: Philosophical Foundations for Populist Reforms*. Berkeley: University of California Press.

Shulman, M. (1990). Bovine Growth Hormone: Who Wins? Who Loses? What's at Stake? In S. Gendel, D. A. Kline, M. D. Warren & F. Yates (Eds.), *Agricultural Bioethics: Implications of Agricultural Biotechnology*. Ames: Iowa University Press.

Singer, B. (1988). An Extension of Rawls' Theory of Justice to Environmental Ethics. *Environmental Ethics, 10*, 217-231.

Sinopoli, R. (1993). Liberalism and Contested Conceptions of the Good: The Limits of Neutrality. *The Journal of Politics, 55*(3), 644-663.

Slovic, P. (1992). Social, Cultural, and Psychological Paradigms. In S. Krismsky & D. Golding (Eds.), *Social Theories of Risk*. Westport: Praeger.

Smith, B. J., & Warland, R. H. (1992). Consumer Responses to Milk from BST-Supplemented Cows. In *Bovine Somatotropin and Emerging Issues*. Boulder: Westview Press.

Sparks, H. (1997). Dissident citizenship: Democratic theory, political courage, and activist women. *Hypatia, 405*(4), 74-110.

StaffReporter. (1986). Judge Rejects Milk Label on Bovine Hormone. *The National Law Journal, 18*, B2.

Stefandies, Z., & Tauer, L. W. (1999). The Empirical Impact of Bovine Somatotropin on a Group of New York Dairy Farms. *American Journal of Agricultural Economics, 81*(1), 95-103.

Steyer, R. (1996, August 11, 1996). Backers and Critics Both Wrong on BST: Small Number of Farmers are Using the Product. *St. Louis Dispatch*, p. IE.
Steyer, R. (1997, March 16, 1997). BST Milk Drug Turns 3 With Little Fanfare: Protests Fade, But Product Lags Expectations. *St. Louis Post-Dispatch*, p. 1E.
Stone, D. (1997). *Policy Paradox: The Art of Political Decision Making*. New York: Norton & Company.
Stout, B. A., & Thompson, P. (Eds.). (1991). *Beyond the Large Farm*. Boulder, CO: Westview Press.
Strauss, L. (1959). *What is Political Philosophy?* Chicago: University of Chicago Press.
Strawley, G. (2001, May 21, 2001). Genetic Engineering arrives in Amish Country. *Associated Press-Indianapolis Star*, p. B7.
Strydom, P. (1999). The Civilization of the Gene: Biotechnological Risk Framed in the Responsibility Discourse. In P. O'Mahony (Ed.), *Nature, Risk and Responsibility*. New York: Routledge.
Tauer, L. (1992). Impact of BST on Small Versus Large Dairy Farms. In M. Hallberg (Ed.), *Bovine Somatotropin and Emerging Issues*. Boulder: Westview Publishing.
Teitel, M., Wilson, K., & Nader, R. (1999). *Genetically Engineered Food: Changing the Nature of Nature*. Rochester: Park Street Press.
Thiele, L. P. (1999). *Environmentalism for a New Millennium*. New York: Oxford University Press.
Thiele, L. P. (2000). Limiting Risks: Environmental Ethics as a Policy Primer. *Policy Studies Journal, 28*(3), 540-564.
Thompson, P. (1990). Biotechnology: Risk, and Political Values: Philosophical Rhetoric and the Structure of Political Debates. In D. Webber (Ed.), *Biotechnology: Assessing Social Impacts and Policy Implications*. New York: Greenwood Press.
Thompson, P. (1992a). Ethical Issues and BST. In M. C. Hallberg (Ed.), *Bovine Somatotropin and Emerging Issues*. Boulder: Westview Press.
Thompson, P. (1992b). *The Ethics of Trade: U.S. Food Policy, Foreign Competition and the Social Contract*. Cambridge: Cambridge University Press.
Thompson, P. (1995). *Spirit of the Soil*. London: Routledge.
Thompson, P. (1997). Food Biotechnology's Challenge to Cultural Integrity and Individual Consent. *Hastings Center Report, 27*(4), 34-38.
Thompson, P. (1998a). *Agricultural Ethics*. Ames: Iowa State University Press.
Thompson, P. (1998b). *Food Biotechnology in Ethical Perspective*. New York: Blackie Academic & Professional.
Thompson, P. (1999). The Ethics of Truth Telling and the Problem of Risk. *Science and Engineering Ethics, 5*(4), 1-22.
Thompson, P. B. (1995). Risk and Responsibility in Modern Agriculture. In T. B. Mephan, G. A. Tucker & J. Wiseman (Eds.), *Issues in Agricultural Bioethics*. Nottingham, NG: Nottingham University Press.
Thompson, P. B., & Dean, W. (1996). Competing Conceptions of Risk. *Risk: Health, Safety and Environment, 7*(Fall), 361-384.
Tolstoy, L. (1967). *On Civil Disobedience and Non-Violence*. New York: Bergman.
Torgerson, D. (1986). Between Knowledge and Politics: Three Faces of Policy Analysis. *Policy Sciences, 19*(1), 33-60.
Torgerson, D. (1997). Policy Professionalism and the Voices of Dissent: the Case of Environmentalism. *Polity, 29*(13), 345-.
Treweek, J. (1996). Ecology and Environmental Impact Assessment. *Journal of Applied Ecology, 33*(2), 191-199.
VonSchomberg, R. (Ed.). (1993). *Science, Politics and Morality*. London: Kluwer Academic Publishers.
Waltzer, M. (1970). *Obligations: Essays on Disobedience, War, and Citizenship*. Cambridge: Harvard University Press.
Warren, K. J. (1999). What Are They Saying about Environmental Ethics? (Book Review). *Ethics, 109*(2), 489.
Warrender, H. (1957). *The Political Philosophy of Hobbes: His Theory of Obligation*. Oxford: The Clarendon Press.
Wenz, P. (1988). *Environmental Justice*. Albany: SUNY Press.
West, G., & Blumberg, R. L. (1990). *Women and Social Protest*. New York: Oxford University Press.
Westra, L. (1998). *Living In Integrity*. New York: Rowman and Littlefield Publishers.

Whipple, C. (1992). Inconsistent Values in Risk Management. In S. Krimsky & D. Golding (Eds.), *Social Theories of Risk*. Westport: Praeger.

Wildavsky, A. (1989). *Speaking Truth to Power*. New Brunswick: Transaction Publishers.

Wildavsky, A. (1993). *Craftways: On the Organization of Scholarly Work*. New Brunswick, NJ: Transaction.

Williams, B., & Matheney, A. (1995). *Democracy, Dialogue and Environmental Disputes: The Contested Languages of Social Regulation*. New Haven: Yale University Press.

Wilson, J. Q. (1967). The Bureaucracy Problem. *The Public Interest, 6*, 193-215.

Wilson, R. (1990). Analyzing the Daily Risks of Life. In T. S. Glickman & M. Gough (Eds.), *Readings in Risk*. Washington, D.C.: Resources for the Future.

Winner, L. (1993). Upon Opening the Black Box and Finding it Empty: Social Constructivism and the Philosophy of Technology. *Science, Technology, and Human Values, 18*(3), 362-378.

Winter, G. (2001, May 31, 2001). Food Industry Agrees to Tighter Standard on Food Labeling. *New York Times*.

Wiser, J. (1983). *Political Philosophy: A History of the Search for Order*. Englewood Cliffs: Prentice-Hall.

Wolff, R. P. (1977). *Understanding Rawls: A Reconstruction and Critique of `A Theory of Justice'*. Princeton: Princeton University Press.

Wooden, C. (2001, July 06, 2001). Food Scientists Agree with Vatican: Technology needs Safeguards. *Catholic News Service*, pp. 1-2.

Wynne, B. (1992). Risk and Social Learning: Reification to Engagement. In S. Krimsky & D. Golding (Eds.), *Social Theories of Risk*. Westport: Praeger.

Young, I. M. (1993). Justice and Communicative Democracy. In R. Gottleib (Ed.), *Radical Philosophy: Tradition, Counter-Tradition, Politics*. Philadelphia: Temple University Press.

Young, I. M. (1995). Communication and the Other: Beyond Deliberative Democracy. In M. Wilson & A. Yeatman (Eds.), *Justice and Identity*. St. Leonards, New South Wales: Allen and Unwin.

Young, I. M. (1997). *Intersecting Voices*. Princeton: Princeton University Press.

Young, R. (1995). Monkeywrenching and the Process of Democracy. *Environmental Politics, 4*(4), 199.

Zashin, E. (1972). *Civil Disobedience and Democracy*. New York: Free Press.

Zietler, A. (2000). *Environmental Democracy: An Analysis of Brownfields*. Unpublished Ph.D., Purdue University, West Lafayette.

Zinn, H. (1972). Disobedience and Democracy. In H. A. Bosmajian (Ed.), *Dissent: Symbolic Behavior and Rhetorical Strategies*. Boston: Allyn and Bacon.

Zwiebach, B. (1975). *Civility and Disobedience*. New York: Cambridge University Press.

The International Library of Environmental, Agricultural and Food Ethics

1. H. Maat: *Science Cultivating Practice.* A History of Agricultural Science in the Netherlands and its Colonies, 1863-1986. 2002 ISBN 1-4020-0113-4
2. M.K. Deblonde: *Economics as a Political Muse.* Philosophical Reflections on the Relevance of Economics for Ecological Policy. 2002 ISBN 1-4020-0165-7
3. J. Keulartz, M. Korthals, M. Schermer, T.E. Swierstra (eds.): *Pragmatist Ethics for a Technological Culture.* 2003 ISBN 1-4020-0987-9
4. N.P. Guehlstorf: *The Political Theories of Risk Analysis.* 2004 ISBN 1-4020-2881-4
5. M. Korthals: *Before Dinner.* Philosophy and Ethics of Food. 2004 ISBN 1-4020-2992-6

KLUWER ACADEMIC PUBLISHERS – DORDRECHT / BOSTON / LONDON

CPSIA information can be obtained
at www.ICGtesting.com
Printed in the USA
LVHW040335050623
748857LV00005B/289